Series editors:
S.A. Corbet and R.H.L. Disney
Advisory board:
J.W.L. Beament, V.K. Brown,
J.A. Hammond, A.E. Stubbs

Published by the Press Syndicate of the University of Cambridge
The Pitt Building, Trumpington Street, Cambridge CB2 1RP
32 East 57th Street, New York, NY 10022, USA
296 Beaconsfield Parade, Middle Park, Melbourne 3206, Australia

© Cambridge University Press 1983
Illustrations © Brenda Jarman 1983

First published 1983

Printed in Great Britain by Belmont Press, Northampton

Library of Congress catalogue card number: 81-38483

British Library Cataloguing in Publication Data
Davis, B.N.K.
Insects on nettles. — (Naturalists' handbooks; 1)
1. Insects — Great Britain 2. Insects —
Identification
I. Title II. Series
595.70941   QL482.G/
ISBN 0 521 23904 4 hard covers
ISBN 0 521 28300 0 paperback

WD

Insects on nettles

B.N.K. DAVIS
Institute of Terrestrial Ecology,
Monks Wood Experimental Station,
Abbots Ripton, Huntingdon

With plates by
BRENDA JARMAN

	Editors' preface and Acknowledgements	page 2
1	Introduction	3
2	Keys to insects on nettles	8
	I. The major groups of insects found on nettles	10
	II. Lepidoptera adults: moths	12
	III. Lepidoptera larvae: caterpillars	13
	IV. Coleoptera: beetles	18
	V. Heteroptera: plant bugs	21
	VI. Homoptera (1): frog-hoppers, leaf-hoppers, etc.	27
	VII. Homoptera (2): plant lice	30
	VIII. Free-living Diptera, Coleoptera and Neuroptera larvae: young stages of flies, beetles and lacewings	32
	IX. Diptera: mines and galls	34
	X. Parasites	36
3	Biology	38
4	Techniques and approaches to original work	50
	Appendix	62
	References and further reading	63
	Index	64
	Glossary	back cover

Plates 1–5 are between pp. 32 and 34

Cambridge University Press

Cambridge
London   New York   New Rochelle
Melbourne   Sydney

Editors' preface

Sixth formers and others without a university training in biology may have the opportunity and inclination to study local natural history but lack the knowledge to do so in a confident and productive way. The books in this series offer them the information and ideas needed to plan an investigation, and the practical guidance needed to carry it out. They draw attention to regions on the frontiers of current knowledge where amateur studies have much to offer. We hope the readers will derive as much satisfaction from their biological explorations as we have done.

The keys are an important feature of the books. Even in Britain, the identification of many groups remains a barrier to ecological research because experts write keys for other experts, and not for general ecologists. The keys in these books are meant to be easy to use. Their usefulness depends very much on the illustrations, the preparation of which was assisted by a grant from the Natural Environment Research Council. S.A.C. & R.H.L.D.

Acknowledgements

I should like to thank all those who have helped me in the production of this book, especially Mrs Brenda Jarman for her patient co-operation in the production of the plates. I have had very useful comments and suggestions on the keys in general from Dr M.G. Morris and Dr R.H.L. Disney and on particular groups from Dr D.J. Carter (Lepidoptera), Dr R.C. Welch (Coleoptera), Dr M.G. Morris (Heteroptera), Dr W.T. Le Quesne (Homoptera: Auchenorhyncha), Dr H.L.G. Stroyan (Homoptera: Aphidoidea) and Dr M.R. Shaw (Hymenoptera: Parasitica); their general form and remaining imperfections, however, are my responsibility. The lists of molluscs and spiders in the Appendix are based on advice from Dr M.P. Kerney and Dr E.A.G. Duffey. The full text has been read by Dr M.G. Morris, Mr H. Berman (St Ivo School) and Mr G. Edwards (Stibbington Field Centre) and again I have received many valuable comments. Mr J.N. Greatorex-Davies has kindly lent specimens and slides of some moths and caterpillars for the illustrations and the checking of the typescript has been painstakingly done by Mr P.E. Jones. To all these gentlemen I am most grateful. I should like also to recall my debt to Mrs Carole Lawrence for her help with studies of the nettle fauna. B.N.K.D.

1 Introduction

There is a natural tendency to avoid Stinging nettles
because of their obvious unpleasant associations. However, I hope this book will convince you that nettles
offer exceptional opportunities for field studies and
simple experiments in ecology. In the first place, nettles
are almost universally available for they grow on clay,
peat and sandy soils, in grassland, woodland, heath and
fen, in agricultural and waste land and around human
habitations; from the coast up to 730 m (2400 ft) in the
Pennines and from the Scilly Isles to the Shetlands. Sec-
ondly, Stinging nettle is a vigorous, long-lived plant
which often occurs in dense stands producing a distinctive habitat which can be examined, treated and
sampled in various ways: unlike many other plants, it
can be readily grown from seed or rhizome (underground stem), or cut down whenever you wish, and its
growth or regrowth can be measured and compared
under different conditions such as the amount of shade
it receives or the kind of soil it grows on. And finally,
Stinging nettle sustains an extremely rich and diverse
insect fauna offering a miniature world for exploration.
Enough is known about these insects to point out some
of their fascinating and contrasting life styles and subtle
interactions, but there is much more we do not yet
know. This book gives keys to the identification of
about 100 species or groups of insects that regularly
occur on nettle, with notes on their biology and
suggested topics for study – but first a few essentials
about the plant itself.

There are two kinds of nettle in Britain, the common,
perennial Stinging nettle *Urtica dioica* L. and the more
local, annual Small nettle *Urtica urens* L. *Urtica* is the
Latin name for nettle whilst *dioica* is derived from the
Greek for two houses (*di-oikos*) and refers to the fact
that this species has separate male and female plants.
Since some insects favour the developing seeds on female
plants and others feed on the pollen of male flowers,
this distinction is important. The flowers appear in pen-
dulous clusters at the tips of the main shoots and larger
side shoots. Though individually small, they are quite
distinctive for the male flowers appear bright yellow
from the pollen on the anthers whereas the styles of the
female flowers have a silvery, furry appearance. Rarely
one may find plants in which the upper part of the
inflorescence bears female flowers and the lower part

male or hermaphrodite flowers and *vice versa*.

Established plants grow rapidly from May till late June and July when they reach their full height of 0.5 to 1.8 m depending on their situation. Giant individuals of over 3 m have been measured! The flowers appear in early June and the ripe seed from July onwards. There is often a second flush of growth in August and September when the side branches bush out and produce more flowers, but thereafter the stems gradually lose their leaves and finally die back with the winter frosts. Fresh growth may start from the base at any time if the stems are cut down, and the lusher foliage is often favoured by caterpillars and other insects. In mild or sheltered areas young shoots may be found even in midwinter. Nettles cannot, however, withstand regular mowing.

The natural situation of *Urtica dioica* is probably open woodland on peaty soils (fen carr) and its present widespread distribution is largely due to the activities of man and his domestic animals. It likes a phosphate-rich soil and often marks former sites of human habitation and accumulations of organic matter such as straw or paper. In poor, sandy areas, such as the Breckland heaths of Norfolk, nettle patches are often associated with rabbit warrens.

Urtica urens stings just as badly as its larger relative (*urens* means burning or stinging) but is distinguished by having its lower leaves shorter than their stalks as well as by the absence of yellow rhizomes and in always having separate male and female flowers on the same plant. It is found on cultivated ground, particularly on light soils. Because it dies and grows from seed every year and rarely forms such dense stands, it does not attract such a large insect fauna.

The stinging hairs of both nettle species may be an effective deterrent against grazing mammals but they appear to offer little defence against caterpillars, which walk over them and eat them with impunity. Occasionally one finds plants of *U. dioica* in which stinging hairs are almost completely absent; this form (var. *subinermis*) is common at Wicken Fen, Cambridgeshire.

All species of plants and animals are given two-part Latin names known as binomials. To avoid any confusion, a binomial is often followed by the name of the author who named and described the species. Both nettle species were named by Linnaeus, who is sometimes called Linné or abbreviated to L.; they are thus definitively cited as *Urtica dioica* Linnaeus or as *Urtica urens* L. The names and authorities for all insect species mentioned in this work are given in the check lists of

British insects (Kloet & Hincks, 1964—78)* together
with alternative names used in earlier works (e.g. the
Peacock butterfly *Inachis io* L. used to be called
Nymphalis io L. and the weevil *Phyllobius pomaceus*
Gyllenhal has been called *Phyllobius urticae* Degeer.)

Species are grouped into genera, genera into families
and families into orders. Thus the genus *Urtica* belongs
to the family Urticaceae which together with the
Cannabaceae and Ulmaceae comprises the order
Urticales. The family tree in fig. 1 sets out all the close
relatives of Stinging nettle which are native in Britain:
Small nettle, Pellitory-of-the-wall *Parietaria diffusa*, Hop
Humulus lupulus and several species of elm *Ulmus* spp.
Though these plants are very different from nettle in
appearance and biology, they must 'smell' or 'taste'
rather similar for there are several insects that feed on
nettles and one or more of these related species but not
on unrelated plants such as White deadnettle *Lamium
album*, which resembles Stinging nettle and often grows
with it. (The parasitic plant Large dodder *Cuscuta
europaea* is also almost confined to nettles and Hop.)

At first sight a bed of nettles may not look as if it is
teeming with animal life since the plants rarely show any
ill effects. A few sweeps with an insect net in high summer will, however, often produce a bewildering variety
of creatures. Ninety per cent or more of these will be
insects but you are also likely to get a number of spiders,
harvestmen, woodlice and snails; these other groups are
not dealt with in this book but the most common snails
and spiders associated with nettle beds are listed in the
Appendix.

Before you can discover what is known or not known
about any insect found on nettles it must be identified
to the order or family to which it belongs and preferably
to the species, which is the basic unit for ecological work.
Eleven orders are described here and it should be possible
at least to place an insect in the correct order with only
the most basic knowledge of insects by using the introductory key and the illustrations. The principal orders
are Coleoptera (beetles), Lepidoptera (moths and butterflies), Diptera (flies), Heteroptera (plant bugs) and
Homoptera (leaf-hoppers, frog-hoppers and aphids, etc.),
which are sometimes loosely combined into the Hemiptera, and parasitic Hymenoptera. (True wasps, bees, ants
and sawflies are not included though they may occasionally be collected.) Each of these orders is then dealt with
in a separate key to adults and/or larvae (caterpillars,

---

* References cited under the authors' names in the text appear in
full in the reference list on p. 63.

6

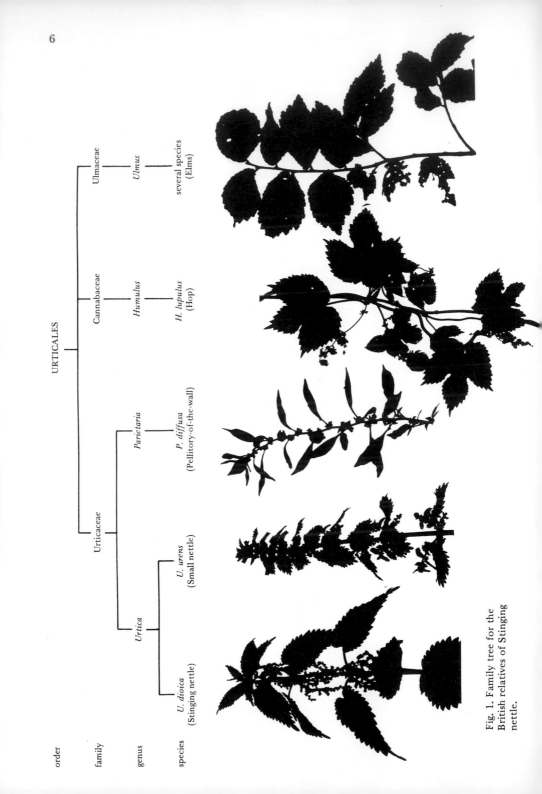

order

family

genus

species

URTICALES

Ulmaceae

Ulmus

several species
(Elms)

Cannabaceae

Humulus

H. lupulus
(Hop)

Urticaceae

Parietaria

P. diffusa
(Pellitory-of-the-wall)

Urtica

U. urens
(Small nettle)

U. dioica
(Stinging nettle)

Fig. 1. Family tree for the
British relatives of Stinging
nettle.

grubs, maggots and nymphs) depending on the stage most often found on nettles; butterflies and moths are rarely found on nettles though they must visit them to lay their eggs.

The five other orders are less important and are not subdivided; in some cases only a single representative is likely to be found. These are Dermaptera (earwigs), Mecoptera (scorpion flies), Neuroptera (lacewings), Orthoptera (crickets) and Thysanoptera (thrips). Table 1 provides a summary of the numbers of insects covered in this book.

Identification beyond this initial stage requires careful study of the specimen concerned. Many of the insects are very small and so some of the characters mentioned can only be seen under a good binocular microscope. The use of technical terms for describing insect anatomy is indispensable but has been kept to a minimum here and all the terms used are illustrated in the keys or described in the Glossary. Concentrate initially on the most common species as these are likely to be part of the true nettle fauna and not just chance visitors. Certain species and groups will soon become familiar and others can then be added more confidently.

Table 1. *Insect orders and numbers of species associated with nettles. 'Restricted' species are those which are more or less confined to Stinging nettle and its close relatives*

| Order | Number of species (and undivided genera) | | | |
	Restricted species	Total species	As adults	As larvae
Coleoptera (beetles)	5	15	15	12
Dermaptera (earwigs)	0	1	1	1
Diptera (flies)	6	7 + Syrphidae (hoverflies)	0	All
Heteroptera (plant bugs)	3	26	26	26
Homoptera (hoppers etc.)	6	23	23	Most
Lepidoptera (butterflies and moths)	10	31	4	31
Mecoptera (scorpion flies)	0	1	1	0
Neuroptera (lacewings)	0	1	1	1
Orthoptera (crickets)	0	1	1	1
Thysanoptera (thrips)	1	1	1	1
	31	107		

2 Keys to insects on nettles

If you do not know its name, the knowledge of it will be lost. (Linnaeus)

General notes
These are artificial keys designed to identify insects, including several common predators, that are known to feed on or have a *regular* association with nettles. Identification is based on scientific names but English names are given where they exist, as in the case of most moths and a few insects that are pests on various crops. In a few cases I have coined English names where it seemed appropriate and unambiguous, e.g. Small nettle aphid, but for many groups there are no common names available. There is also no general English name to cover all the insects within the order Homoptera.

Many other insects which are not included in the keys will be collected from nettles; these are probably just sheltering amongst the dense vegetation or they may be insects which are really associated with other low plants or trees growing nearby. It is impossible to allow for all these 'non-nettle' insects without greatly extending the keys and including diagnostic characters that are difficult to use without considerable experience. *You must therefore be prepared to reject a specimen as being merely a casual visitor on nettles* if it does not fit any of the possibilities given, though the keys may allow you to place it in a family or genus. To guard against misidentifications, you should check all the details given for a species. (Note: obvious features are not always distinctive ones as far as keys are concerned because they may be shared by several species.)

The keys themselves are set out in numbered pairs of alternatives called couplets. Choose the alternative that fits your specimen and go on to the number indicated. If you appear to have 'gone wrong' you may wish to work backwards through the keys; numbers in brackets indicate the couplet to go back to when it is not the immediately preceding one. **Use the illustrations to check the descriptions.** Keys V—VII and X (plant bugs, hoppers, aphids, etc. and parasites) include many small insects requiring strong magnification (at least ×20) and illumination to see some of the characters referred to. These groups also involve the use of more technical terms. Beginners should therefore start with beetles (key IV) or caterpillars (key III).

Segments. Insect bodies and limbs are divided into segments which often have special names. Abdominal segments are numbered from front to back; leg and

G.1. *Macrosteles* sp. ♀
(Homoptera)

G.2. *Liocoris tripustulatus* ♀
(Heteroptera)

G.3. *Macrosteles* sp. ♂
(Homoptera)

G.4. *Liocoris tripustulatus* ♂
(Heteroptera)

Symbols and abbreviations

* species more or less restricted to nettles and close relatives
♂ male
♀ female
sp. species
spp. species (plural)

antennal segments are numbered from the end nearest the body. Likewise, the base of a segment is the end nearest the body and the apex or apical end is furthest from the body.

Size. The size of species is often an important character and a pair of fine, adjustable dividers should be used for measuring. Lengths given are of the whole insect from front of head to end of abdomen, or wings if these extend beyond the abdomen, unless otherwise stated. Specimens preserved in alcohol may become distorted by expansion of the joints.

Colours are of fresh, unrubbed specimens. They may fade in pinned specimens or become apparently paler and translucent (especially on wings) in alcohol.

Seasonal occurrence. It is often helpful to know when the different species can be found on nettles. Species names are therefore preceded by the months when they occur as adults, unless the key is specifically concerned with larvae (e.g. caterpillars). 'H' indicates hibernation and '—' indicates overlapping broods. Thus: June—Aug & (Aug)Sept—H—March—May means that a first generation is found from June to August and that a second generation appears occasionally in August but mainly in September before hibernation and then again in the spring from March to May. (July—Aug) means that a second generation occurs in some years in July and August. Jan—Dec means that broods occur throughout the year from January to December.

Sex differences. It is sometimes necessary to distinguish males from females, e.g. amongst the Homoptera and Heteroptera, for purposes of identification. The elongate, horny and often strongly curved genital valves on the underside of the female are easily recognised in both these groups (figs. G.1, G.2). Male Homoptera have paired genital plates which differ in form considerably between families, while male Heteroptera show little or none of the genital structures externally (figs. G.3, G.4). Female parasitic Hymenoptera are recognised by their sting (ovipositor) at the tip of the abdomen.

A more comprehensive and well-illustrated general guide to insect groups is provided by Chinery's (1976) *A Field Guide to the Insects of Britain and Northern Europe*, but many nettle insects are not described.

References to other keys or works dealing with specific groups are given at the end of each of the following keys.

I The major groups of insects found on nettles

I.1. Earwig *Forficula auricularia* ♀

1 Insects found on nettles   2

— Parasites hatching out of nettle insects   key X (p. 36)

2 Free-living insects found on the surface of the plant   3

— Grubs, maggots, etc., within the plant tissues making mines and galls (and the adults produced by them)
key IX (p. 34)

3 Winged insects. Wings often meeting in the mid-line or overlapping; sometimes horny or very short (I.1), or like lobes or flaps in the case of young stages (V.9)   4

— Entirely wingless insects (caterpillars, some aphids, etc.)   15

4 Wings covered with coloured scales. MOTHS
key II (p. 12)

I.2. *Thrips urticae*

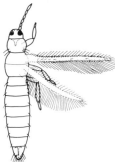

— Wings without scales   5

5 Hind legs very much larger than the others (adapted for jumping). CRICKETS etc. One species on nettles (pl. 5.3). Adults up to 20 mm. (Apr)May—Sept(Nov) (nymph—adult)
     Dark bush-cricket *Pholidoptera griseoaptera* (see p. 42)

— Not as above   6

6 Body ending in a pair of large forceps (I.1). EARWIGS. One species on nettle. Up to 14 mm
     Common earwig *Forficula auricularia* (see p. 42)

— Not as above   7

I.3. Beetle (Carabidae)

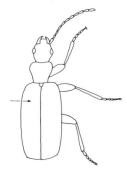

7 Minute, narrow, yellowish insect 1—1.3 mm long. Fringed wings visible under high magnification (I.2). 'Thunder-bugs', THRIPS. One species restricted to nettles
May—Oct   *Thrips urticae**

— Not as above   8

8 With a hard or leathery pair of wing cases (elytra) meeting down the middle of the back (I.3); sometimes very short (IV.6). (Without veins though sometimes with visible striations.) BEETLES   key IV (p. 18)

— Without wing cases. Forewings usually clearly veined   9

I.4. Hoverfly *Syrphus ribesii*

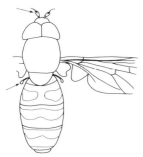

9 With only 1 pair of delicate wings (usually held out flat and directed obliquely backwards or overlapping flat over the back) and a pair of small club-like balancing organs (halteres) at their bases (I.4). (Head usually with a large pair of eyes and short antennae (I.4) or with longer, jointed, feathery antennae.) FLIES and MIDGES
key IXB (p. 34)

— With 2 pairs of wings, the hind pair often small and hidden under the forewings. (If the wings are held out or overlapping flat over the back the antennae are longer than the head and without whorls of hairs) 10

I.5. Scorpion fly *Panorpa* sp.
♂

10 Transparent wings folded tent-like over the body; with very many veins producing a ladder-like effect along the front (lower when folded) edges (pl. 5.4). Body with wings up to about 16 mm
Jan—Dec LACEWINGS (see p. 41)

— Not as above 11

I.6. Scorpion fly *Panorpa* sp.
♂

11 Face prolonged downwards into a distinctive beak (I.5). Transparent wings with several dark spots, held out flat and directed obliquely backwards. Slender abdomen turned up like a scorpion's tail in the male (I.6). Wings 12–15 mm
May–Aug SCORPION FLIES *Panorpa* spp. (see p. 42)

— Head and abdomen not as above. Wings usually folded flat or tent-like over the back 12

I.7. Homoptera *Aphrophora* sp.

12 Wings folded tent-like over the back (I.7). With piercing and sucking mouthparts visible in side view (I.8), sometimes appearing between the legs (I.11). Small insects often less than 5 mm. HOMOPTERA 13

— Not as above 14

I.8. Homoptera *Macrosteles* sp.

13 Antennae of 1 or 2 short segments and a terminal bristle (I.8). FROG-HOPPERS, LEAF-HOPPERS etc.
key VI (p. 27)

— Segmented antennae much longer than head. With a pair of prominent cones on the forehead (I.9) or a pair of conspicuous tubes (siphunculi) (I.11) on the abdomen. Small delicate insects 1—4 mm. PLANT LICE (APHIDS and PSYLLIDS) key VII (p. 30)

I.9. Psyllid *Trioza urticae*

I.10. Plant bug *Nabis* sp.
(Heteroptera)

I.11. Aphid *Microlophium carnosum*

I.12. Caterpillar *Orthosia* sp.

spiracle

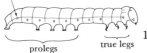

prolegs   true legs

See:
Chinery (1976)
Lewis & Taylor (1967)
Oldroyd (1958)
Paviour-Smith & Whittaker
(1967)

14 Forewings folded flat over the back; with a leathery basal part and delicate tips (pl. 4). Segmented antennae longer than head. Long, piercing and sucking mouthparts, often tucked back under the head (I.10) and sometimes visible only from below. 3—11 mm. PLANT BUGS   key V (p. 21)

— Not as above. (This will include several different groups of insects such as ants, bees, wasps and sawflies (order Hymenoptera), or stoneflies, mayflies and caddis flies (if near water), none of which has any special association with nettles)

15(3) With sucking mouthparts and long antennae (I.10, I.11). Small, delicate insects   16

— Not as above; caterpillars, grubs, maggots, etc.   17

16 With a pair of conspicuous tubes (siphunculi) on the abdomen (I.11). Wingless APHIDS   key VII (p. 30)

— Without siphunculi. Youngest stages of PLANT BUGS   key VB (p. 25)

17 Caterpillars; with well-formed head capsule, 3 pairs of jointed true legs and 2—5 pairs of prolegs (I.12). BUTTERFLY and MOTH LARVAE   key III (p. 13)

— Without this combination of characters. FLY, BEETLE and LACEWING LARVAE   key VIII (p. 32)

II Lepidoptera adults: moths

Four species of moth, whose larvae are more or less restricted to feeding on nettles, may be readily disturbed from patches of nettle at the appropriate season and are therefore keyed below. Most of the other nettle-feeding Lepidoptera are seldom seen on nettles as adults. Adults bred from larvae should be identified from illustrations in books listed in the references below.

1 Very small, dark brown moth with a wing span of about 13 mm but looking like a little dart when the wings are folded back in the natural resting position (pl. 1.5) May—June & July—Sept

Nettle-tap *Anthophila fabriciana**

— Larger moths with outstretched wing span of about
30—35 mm   2

2 Plain brown moth with a pair of feathery tufts projecting forwards from the front of the head (pl. 1.6)
June—July   Snout *Hypena proboscidalis**

— Pale-coloured moths with darker mottling   3

3 Cream-coloured with beige mottling (pl. 1.2)
July   Mother-of-pearl *Pleuroptya ruralis**

See:
Beirne (1952)
South (1963)

— Whitish wings with strongly contrasting dark brown
edging and spots. Tip of the abdomen pale orange
June—July   Small magpie *Eurrhypara hortulata*

III   Lepidoptera larvae: caterpillars

III.1. *Aglais urticae*
(Nymphalidae)

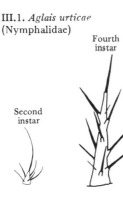

Fourth
instar

Second
instar

Many caterpillars can be recognised at a glance with
experience but the variations that can occur between
different growth stages (instars) of an individual caterpillar and among different individuals of the same
species make this a difficult group to identify from
descriptions. *The characters given refer mainly to the
fully grown larvae* and some species must be bred out
to the adult for certain identification. Sawfly larvae
(Hymenoptera) may occasionally be collected from
plants growing with nettles. These look like caterpillars
but can be easily recognised by having six or more pairs
of stumpy prolegs whereas Lepidoptera caterpillars
have a maximum of five including the terminal pair
(I.12).

1 Caterpillars with branched spines (III.1) on each segment
(except in the youngest stages); sometimes in a communal web (butterfly larvae)   2

— Caterpillars hairy or hairless but not spiny and never in a
communal web (moth larvae)   5

Nymphalidae

2 Bicoloured; segments 1—5 dark- or reddish-brown, 6—11
white on the back with white spines (pl. 2.4)
May—June & (July—Aug)   Comma *Polygonia c-album**

— Uniformly patterned; generally dark with pale dots and
lines   3

3 Velvety black caterpillar with black spines and sprinkled with white dots. Younger stages in colonies in a conspicuous web
May/June—July   Peacock *Inachis io**

— Caterpillars with pale ground colour or pale spines or lines   4

4 Typically black with considerable development of yellow lines and dots and yellow or black spines. Young stages colonial in a web but becoming scattered and solitary. (Older individuals make a tent when about to moult)
May—June & July—Aug
                     Small tortoiseshell *Aglais urticae**

— Greenish, brownish or blackish caterpillar with yellowish or black spines and variously freckled with white or yellow. Solitary. Concealed within a tent-like chamber of one or more leaves drawn together
(June—July) & Aug—Sept
                     Red admiral *Vanessa atalanta**

Moth larvae

5(1) Caterpillars feeding within an individual web or rolled leaf or between leaves spun together   6

— Caterpillars fully exposed. (Caution: if nettles are shaken or swept some of the above species may be collected and if their association with a rolled leaf or web is not realised they may be wrongly sought here.)   12

6 Small species living under a web spun usually on the upper surface of a leaf and often with the edges of the leaf drawn together. Pale creamy-green or greenish-white caterpillar up to 12 mm, with dark spots on each segment (III.2)
June—July & Aug—H—May
                     Nettle-tap *Anthophila fabriciana**

III.2. *Anthophila fabriciana*

— Leaves variously rolled or spun together but without a conspicuous web   7

7 Small, chocolate-brown or black caterpillars within a silk-lined cell in a folded leaf. (Leaf often wilted and hanging down when the stalk has been partly eaten through. Larvae very active when disturbed.)
(Tortricidae)   8

— Greenish, whitish or pinkish caterpillars. (Pyralidae)   9

Tortricidae

8 Older stages developing pale spots and side lines
May—June   *Clepsis spectrana*

— Without pale spots and side lines
Aug—H—May   *Olethreutes lacunana*

(These two species may be confused with other closely related species which also feed on a wide range of plants and are occasionally found on nettle. Breed out to confirm identity.)

Pyralidae

9(7) Blue-green or glossy-green caterpillar without black spots or pale stripes. (Young stages with grey spots.) Young stages roll the edge of a leaf, but older stages roll the whole leaf into a tube *with a few strong strands of silk* (pl. 2.9)
Sept—H—June   Mother-of-pearl *Pleuroptya ruralis**

— Leaf-rolling caterpillars with large black spots or pale stripes   10

10 Transparent green or blackish-green caterpillar with several large intensely black spots on each segment, each bearing a short black hair
Sept—H—Apr   *Udea olivalis*

— Caterpillars without black spots   11

11 Bright transparent green caterpillar with a pair of broad opaque white stripes down the back and pale warts with short white hairs
Oct—H—May   *Udea prunalis*

— Whitish or pale lemon-green caterpillar with a pair of pale lines down the back and a black head. (Mature larvae become flesh-pink in the autumn before hibernation)
Aug—H—May   Small magpie *Eurrhypara hortulata*

12(5) Very hairy caterpillars, hairs arising in tufts (Arctiidae). (Several species feed on a wide variety of low-growing plants; the 6 most commonly found on nettle are keyed out)   13

— Caterpillars with few or no conspicuous hairs (Noctuidae)   18

Arctiidae

13 Caterpillar with crimson-red head, legs and prolegs.
    Blackish body with clusters of reddish-brown hairs
    July—Aug—H—March   Cream-spot tiger *Arctia villica*

— Without red head and legs   14

14 Caterpillar black, with yellow bands or more or less
    connected yellow spots down the middle of the back
    and along the sides. Rather short grey and black hairs
    July—Aug—H—May   Scarlet tiger *Callimorpha dominula*

— Without yellow spots   15

15 Black caterpillar with very long black hairs. In full-grown
    specimens these are whitish at their tips and reddish-
    orange around the head and flanks. 'Woolly bear'
    Aug—Sept(Oct)—H—June   Garden tiger *Arctia caja*

— Brown or grey caterpillars with shorter reddish-brown or
    grey hairs   16

16 Caterpillar with a pale or dark line down each side
    separating a dark back from paler brown or greyish sides
    (pl. 2.8)
    Aug—Sept/Oct   Buff ermine *Spilosoma luteum*

— Without dark back and paler sides   17

17 Head pale chestnut-brown (yellow when young). Dark
    brown caterpillar with short golden-brown or greyish
    hairs arising from pale warts
    July—Aug/Sept   Muslin *Cycnia mendica*

— Head black. Dark brown or grey caterpillar with dense
    tufts of blackish hairs
    Sept—H—May & (June—July)
                         Ruby tiger *Phragmatobia fuliginosa*

Noctuidae

18(12) With only 3 pairs of prolegs (including terminal pair).
        (Plusiinae). A group of very similar and variable cater-
        pillars ranging from bright or yellowish-green to very
        dark olive-green with longitudinal lighter or darker lines.
        Breed out to identify   19

— With 4 or 5 pairs of prolegs (including terminal pair)   20

19 Caterpillar without black streak along side of head
    (pl. 1.7)
    Sept—H—May & (July—Aug)
                     Burnished brass *Diachrysia chrysitis**

— Species with black streak along side of head
May—Sept   Silver Y *Autographa gamma*
July/Aug—H—May
           Beautiful golden Y *Autographa pulchrina*
July/Aug & Sept—H—May
              Plain golden Y *Autographa jota*
Aug—H—May/June   Gold spangle *Autographa bractea*

20(18) With 4 pairs of prolegs. Rather slender nettle-green
caterpillar with obvious constrictions between segments
and moderately long dark hairs (pl. 2.7)
Aug—H—May   Snout *Hypena proboscidalis**

— With 5 pairs of prolegs. (Several other noctuids feed on a
wide variety of low-growing plants. The 7 most commonly found on nettle are included below with two that
are confined to nettle)   21

III.3. *Orthosia gothica*

21 Body tapering at rear end and perfectly smooth from
head to tail (III.3)   22

— Eighth abdominal segment (eleventh after the head)
distinctly angled or humped. Abdominal segments 1 and
2 sometimes slightly humped (III.4)   25

III.4. *Melanchra persicariae*

22 Green caterpillars with continuous fine white or yellowish lines down the middle of the back and on the upper
sides, as well as a broader spiracular stripe (III.3). Without disjointed light or dark markings on segments. (Two
very similar species)   23

Caterpillars without continuous pale mid-dorsal and
lateral lines apart from spiracular line   24

23 Larger species, up to about 4 cm. Greyish-green often
with prominent dark line above pale spiracular line
(pl. 2.2)
Apr/May—June   Hebrew character *Orthosia gothica*

— Smaller species, up to about 3 cm. Bright green
June—July   Mouse *Amphipyra tragopoginis*

24(22) Slim tapering species. Green to pale brown with yellow
spiracular stripe often darkly bordered above. Minutely
dotted with white and black dots
July—Sept   Bright-line brown-eye *Lacanobia oleracea*
(The much less common Plain clay *Eugnorisma depuncta*
may key out here but is usually distinguishable by a
series of dark diamonds or crosses along the back.)

— Soft, plump, mottled yellowish-green or brownish-green
species with obscure dusky Vs along the back
Jan——Dec   Angle shades *Phlogophora meticulosa*

III.5. *Abrostola triplasia*

Third instar   Fifth instar

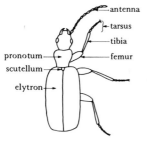

25(21) Light green caterpillar with series of overlapping white Vs down the back (III.5). In side view a series of white and darker green lines are seen running diagonally backwards both above and below a white side line (pl. 1.3). (A brown form can occur)
July & (Sept)   Spectacle *Abrostola triplasia**

— Green or brown caterpillars with strong dark markings on abdominal segments 1, 2 and 8. (Three similar species)   26

26 Rather slender, body tapered from middle to head. Protuberance on abdominal segment 8 with 2 conical points. Pale spiracular line and white-edged, dark, oblique lines on the sides of abdominal segments 3—8
July/Aug—Sept   Dark spectacle *Abrostola trigemina**

— Larva stout, not tapered from middle of body to head. Abdominal segment 8 without conical points   27

27 Dark brown or green plate on first thoracic segment (behind head) with 3 distinct, longitudinal, white lines. Spiracular line not conspicuously expanded on abdominal segment 8. V-shaped dark markings on the back (pl. 2.3)
Aug—Oct   Dot *Melanchra persicariae*

See:
Carter (1979)
Stokoe & Stovin (1958)

— First thoracic plate dark brown with paler marbling. Pale spiracular line broadly expanded to form a large triangular patch on the side of abdominal segment 8
July—Oct   Flame *Axylia putris*

IV Coleoptera: beetles

IV.1. Carabid beetle

antenna

tarsus

tibia

pronotum — femur

scutellum

elytron

A large group of insects, very variable in size and form but with parts of the body as in IV.1.

1 Weevils; with characteristic long snout (rostrum) (IV.2) bearing elbowed antennae except in *Apion* spp. (IV.3) (Rostrum sometimes tucked down between legs)   2

— Other beetles   5

Curculionidae and Apionidae (weevils)

2 Black or grey, ovoid species with very long downward curving rostrum. 2—4 mm   3

— Large green or tiny reddish species; rostrum straight 4

IV.2. *Phyllobius pomaceus*

IV.3. *Apion urticarium*

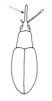

3 Larger species. Front femur strongly toothed beneath and tibiae black. Without white patches on elytra. 3—3.7 mm (pl. 3.2)
Aug—Oct—H—Apr—Jun(Aug)
*Ceutorhynchus pollinarius**

— Smaller species. Front femur not or very weakly toothed. Tibiae reddish. A patch of white scales in the middle of each elytron at the sides and a patch at the tip of the elytra near the mid-line, sometimes indistinct. 2—3(—3.5) mm. Very common
Aug—Nov—H—Apr—June(Aug)
*Cidnorhinus quadrimaculatus**

4(2) Large green weevil (older individuals sometimes appearing black when scales are rubbed off). Rostrum rather short and broad. 7—9 mm (pl. 3.1)
May—June(Aug) *Phyllobius pomaceus**

— Very small, narrow species. Dark red with whitish scales. Antennae arising from base of rostrum. 1.6—2.3 mm (IV.3)
Aug—Oct—H—June—July *Apion urticarium**
(Many other *Apion* spp. may be collected but have no association with nettles.)

5(1) Ladybirds; yellow or red, hemispherical beetles with black spots. (All-red to all-black colour forms occur in some species. Only the 5 most common species are keyed out here) 6

— Other beetles 10

Coccinellidae (ladybirds)

6 Large, 5.5—7.5 mm. Elytra usually red with 7 black spots including a median shared scutellary spot. Pronotum black with white patches in anterior (front) corners. (Varieties occur in which the elytra range from all-red to all-black)
Seven-spot ladybird *Coccinella 7-punctata*

— Smaller, 3—4.5 mm 7

IV.4. *Coccinella 11-punctata*

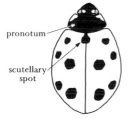

pronotum

scutellary spot

7 Elytra red with 11 black spots including a median scutellary spot (IV.4). (Pronotum with small yellow patches in anterior corners. Legs and undersides black.) 3.5—4.5 mm
Eleven-spot ladybird *Coccinella 11-punctata*

— Without this clear pattern of spots 8

8 Legs and undersides black. The typical form has red elytra with 1 black spot on each. Other varieties have

IV.5. *Adalia bipunctata* var.
sexpustulata

black elytra with 1—3 pairs of red spots and either red patches at the shoulders (IV.5) or a red band right across the base of the elytra. 3—4.5 mm

Two-spot ladybird *Adalia bipunctata*

— Legs and underside partly yellow   9

9 Elytra yellowish with large rectangular black spots, those near the mid-line usually joined to a black median streak. 3—4 mm (pl. 5.1)

Fourteen-spot ladybird *Propylea 14-punctata*

— Elytra without rectangular black spots; varying from entirely reddish-yellow to nearly all black, but typically with 5 pairs of black spots. Pronotum equally variable in marking. 3—4 mm

Ten-spot ladybird *Adalia 10-punctata*

Other families

10(5) Very small blackish beetles 1.5—2 mm, often swarming on male flowers. Tip of abdomen projecting beyond blunt elytra. Short antennae with clubbed ends   11

— Larger beetles with long simple antennae   12

11 Entirely black or with femora dull red. Longish silvery hairs on elytra
May—Aug   Flower beetle *Brachypterus glaber*

— Dark pitchy-red with reddish legs and antennae. Shorter yellowish hairs on elytra (pl. 3.3)
May—Aug   Flower beetle *Brachypterus urticae**

IV.6. *Tachyporus* sp.

12(10) With short, waistcoat-like elytra exposing most of the abdomen. (Rove-beetles: Staphylinidae)   13

— With long elytra covering all or most of the abdomen   14

13 Hind body strongly tapering (segments may be telescoped, each inside the one in front) with long dark setae along side margins of thorax, elytra and hind body. Head small, antennae slender (IV.6). Mostly shiny, black and/or yellowish-red beetles. 2.5—4 mm

Rove-beetles *Tachyporus* spp.

— Without this combination of characters
Other Staphylinidae

IV.7. *Crepidodera ferruginea*

14(12) Third tarsal segment two-lobed. Middle of hind femur broad and fat, for jumping. Base of thorax with a depressed section between longitudinal grooves (IV.7). Entirely reddish-yellow, ovoid beetle. 3—4 mm

Flea beetle *Crepidodera ferruginea*

IV.8. *Demetrias atricapillus*

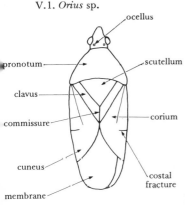

— Not as above. More elongate beetles with long slender legs 15

15 Front tibiae notched on inner side near tip (IV.8). Fourth tarsal segment deeply two-lobed. Head black, elytra reddish, sometimes with a dark streak down the mid-line gradually broadening towards rear end. 4.5—5.6 mm (pl. 3.4). (Similar, related species may occasionally be found but they do not have this combination of characters)
Aug—Nov—H—Apr—May   *Demetrias atricapillus*

— Front tibiae simple 16

16 Elytra rather soft, more or less parallel-sided. Fourth tarsal segment two-lobed. (Cantharidae) 17

— Not as above
Other beetles, e.g. Click beetles (Elateridae)

17 Joints 1—3 of tarsi simple. Entirely reddish-yellow except for black tarsi and antennae (apart from first joint) and ends of elytra. 8—10 mm (pl. 5.2)
Soldier beetle *Rhagonycha fulva*

See:
Joy (1932)
Linssen (1959)

— Not as above
Other Cantharidae

V   Heteroptera: plant bugs

V.1. *Orius* sp.

ocellus

pronotum

scutellum

clavus

commissure

corium

cuneus

costal fracture

membrane

This is a large and rather difficult group with many common species that may be encountered when sweeping nettles that are growing under trees or amongst mixed herbage. The five subfamilies of the family Miridae are largely characterised by differences in the accessory claw structures and these have not been used here because they are often so difficult to see. It is therefore particularly important to check that all the given characters fit the specimen that is being identified. Some of the more common 'non-nettle' plant bugs are given at the end against the species in the key with which they could be confused.

A. ADULTS

1 Rostrum (mouthparts) curved when viewed from side (I.10). Large species (more than 6 mm). Wings without costal fracture and cuneus, usually shorter than abdomen. (Nabidae) 4

— Not as above 2

2   Rostrum with 3 segments, not pressed against underside
when at rest. (Forewings with costal fracture and cuneus
(V.1). Head and thorax dark brown or black. With ocelli.
Small species (less than 4.5 mm).) (Cimicidae)   6

—   Rostrum straight and pressed against body when at rest,
4-segmented   3

V.2. *Scolopostethus* sp.

3   Forewings without a costal fracture and cuneus.
Scutellum equal to or more than length of commissure
(V.2). (With ocelli. Black or brown bugs with rather
short antennae.) (Lygaeidae)   8

—   Forewings with costal fracture and cuneus (except *O.
rufifrons* ♀ with reduced wings). Scutellum shorter than
commissure. (Ocelli absent. Very variable in size and
shape. Colour often green, yellow, brown, red, grey or
black. Often with long legs and antennae.) (Miridae)   10

Nabidae

4(1) Abdomen with yellow or orange spots around the
margin (connexivum). Usually brachypterous, i.e. wings
slightly shorter than abdomen (pl. 4.5). 7—8 mm
(July)Aug—H—Apr—July

Ant damsel bug *Aptus mirmicoides*

—   Connexivum without spots   5

5   Larger species. Usually micropterous, i.e. wings only
covering second and part of third abdominal segment.
7.5—9 mm
July—Nov   Marsh damsel bug *Dolichonabis limbatus*

—   Smaller species. Usually fully winged or brachypterous.
(If micropterous, then hind margin of pronotum about
2½ times as wide as front margin.) 6.4—7.6 mm
Aug—H—May—July   Common damsel bug *Nabis rugosus*

Cimicidae

V.3. *Anthocoris* sp.

6(2) Very small species, mostly less than 3 mm. Oval in out-
line: at most 2½ times as long as broad. Scutellum longer
than commissure (V.1)
June—Aug & (Aug)Sept—H—Mar—May   *Orius* spp.

—   Larger species, more than 3 mm. Narrowly oval: about
3 times as long as broad. Scutellum and commissure
about equal in length (V.3)   7

7   Forewings entirely shining, light-coloured with brown
and black areas including brown tips. Pronotum entirely
black. 3.4—4.1 mm (pl. 5.6). Very common
(June)July—Sept & Sept—H—Mar—May
Anthocoris nemorum

— Forewings partly dull, not patterned as above
Anthocoris nemoralis and other spp.

Lygaeidae

8(3) Larger species. Fully winged. Head, thorax and legs with long, erect hairs (pl. 4.4). 6—7 mm
Sept—Oct—H—May—July
Nettle ground bug *Heterogaster urticae**

— Smaller species. Usually short-winged (brachypterous) (V.2). Hairs inconspicuous   9

V.4. *Scolopostethus affinis* ♂

9 Male (fig. G.4, p. 9) with a pair of curved processes (V.4) — female (fig. G.2, p. 9) with short tubercles — between bases of first and second pairs of legs. First, second and sometimes base of third antennal segment pale. 3.3—3.4 mm
Sept—H—June—July   *Scolopostethus affinis*

— Male and female without processes or tubercles between bases of first and second legs. Second antennal segment usually dark at its end. 3.5—4.1 mm
Sept—H—June—July   *Scolopostethus thomsoni*

Miridae: capsids

V.5. *Lygus* sp.

10(3) Thorax constricted in front to form a collar (V.5)   11

— Thorax not constricted   22

11 Large, broadly oval species marked with varying amounts of light brown and black and at least the cuneus partly red (pl. 4.7). Claws with large tooth like

V.6. *Deraeocoris* sp.

process at base (V.6). 6.5—7.5 mm
July—Oct   *Deraeocoris ruber*

— Not as above   12

12 Head and thorax green. Wings all green or suffused with red or smoky-brown   13

— Head and thorax not green, and often brown or black   17

13 Large species. Entirely pale green. 10—11 mm
June—July   *Calocoris alpestris*

— Smaller species. 5—8 mm   14

14 Hairs on pronotum, wings and tibiae black. Wings sometimes suffused with red or smoky-brown. 6—8 mm (June)July—Aug(Oct)
Potato capsid *Calocoris norvegicus*

— Hairs on pronotum and wings pale   15

15 Hairs on tibiae fine and pale brown (pl. 4.6). 5—6.6 mm
June—July & Aug—Oct
Common green capsid *Lygocoris pabulinus*

— Hairs on tibiae coarse and black (not arising from black
spots) 16

16 Cuneus black at tip. 5.3—6 mm
(June)July—Sept *Lygocoris spinolai*

— Cuneus entirely green. 5—6 mm
July—Oct *Lygocoris lucorum*

17(12) With a transverse ridge between the eyes at the back of
the head (vertex) (V.5) 18

— Vertex without a transverse ridge 19

18 Upper surface of body and wings densely clothed with
short, fine hairs. Corium not very shiny. Colour varying
from almost entirely black to pale grey. 4.8—5.7 mm
July—Aug & Sept—H—Mar—May
European tarnished plant-bug *Lygus rugulipennis*

— Hairs inconspicuous so that insect appears shiny. Dark
red to pale grey-brown. Northern and upland species.
5.4—6.8 mm
Aug—H—Apr—June *Lygus wagneri*

19(17) Smaller species, up to 5 mm. Black spines on hind tibiae
as long as width of tibiae 20

— Larger species, over 5.5 mm. Spines on hind tibiae distinctly shorter than width of tibiae 21

20 Broadly oval species. Head and pronotum largely brown
or black, scutellum yellow, cuneus with broad pale band
and black tip (pl. 4.3). 3.8—5 mm. Very common
July—H—Apr—July
Common nettle capsid *Liocoris tripustulatus**

— Narrow, elongate species. Head and pronotum more or
less black, forewings dark grey. (Legs green with black
spots. Antennae mainly dark.) 4.5—5 mm
June—Oct(?H) *Dicyphus errans*

21(19) Forewings largely dark brown, with gold scale-like hairs.
5.8—7 mm
June—Aug(H) *Calocoris fulvomaculatus*

— Forewings largely black with 2 yellow or light green
patches and cuneus orange. Without scale-like hairs.
5.6—7.6 mm
June—Aug *Calocoris sexguttatus*

22(10) Dark second antennal segments much thickened and flattened (about 5 times as wide as segment 3). (Head, pronotum and wings black or dark brown with long, outstanding hairs. Legs entirely pale.) (pl. 4.1).
4.6–5.5 mm
July–Sept(Oct)   *Heterotoma planicornis*

— Antennae not as above   23

V.7. *Plagiognathus arbustorum*

23 Spines on hind tibiae pale and short. Legs without dark spots   24

— Spines on hind tibiae at least as long as width of tibiae and arising from dark spots. Hind femora with dark spots (V.7). (Black hairs on thorax and wings)   25

V.8. *Orthotylus ochrotrichus*

24 Head, pronotum, wings and antennae bright green, sometimes with yellow markings. Hairs pale. (Basal segment of antennae shorter than head and with 4 long hairs, 3 visible dorsally.) (V. 8). 4.4–5.1 mm
July–Sept   *Orthotylus ochrotrichus*

— Male narrow, fully winged, black or grey; female broadly oval, brachypterous, black with red head (pl. 4.2). First 2 segments of antennae black in male; at least apex of second segment black in female, remainder pale. Legs pale yellow. ♀ 3–3.5 mm, ♂ 4–4.5 mm
July–Sept   *Orthonotus rufifrons**

25(23) Head, forewings, thorax and abdomen partially green. Femora without dark streaks. 3.3–4.1 mm
(June)July–Sept   *Plagiognathus chrysanthemi*

— Head, forewings, thorax and abdomen vary from light red-brown to almost black, and without any green. Dark streaks on fore and rear edges of mid and hind femora (V.7). 4–4.5 mm. Very common
July–Sept   *Plagiognathus arbustorum*

See:
Southwood & Leston (1959)

V.9. *Scolopostethus* sp. (Lygaeidae)

B. FAMILIES OF HETEROPTERA LARVAE

1 Three dorsal abdominal glands (paired or single) situated in dark patches behind segments 3–4 (V.9)   2

— Less than 3 dorsal abdominal glands though dark patches may be present on 5 segments   4

V.10. *Heterogaster* (Lygaeidae)

2 Rostrum straight, pressed to body when at rest. 4-segmented   (*Scolopostethus*) Lygaeidae

— Rostrum more or less curved (I.10), not pressed to body when at rest   3

V.11. *Liocoris* (Miridae)

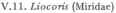

See:
Southwood (1956)
Southwood & Scudder (1956)

3 Forelegs powerful (for grasping prey), tibia fitting closely to femur. Rostrum 4-segmented   Nabidae

— Forelegs not like this. Rostrum 3-segmented. Small species bright red or brown   Cimicidae

4(1) Two dorsal abdominal glands, opening behind segments 4 and 5 (V.10)   (*Heterogaster*) Lygaeidae

— One dorsal abdominal gland (V.11)   Miridae

POSSIBLE MISIDENTIFICATIONS

Deraeocoris ruber.   *D. lutescens* is distinguished by its smaller size (less than 5 mm) and being found as adult in May and June.

Dicyphus errans.   *D. constrictus* and *D. epilobii* differ in having a green head, pronotum and first and second antennal segments (yellowish-grey after death). Occur on glandular/hairy herbaceous plants.

Orthotylus ochrotrichus.   Very similar to *O. prasinus.* The latter is associated with elm, oak, etc., and the vertex is usually keeled.

Calocoris norvegicus
Lygocoris spp.
Lygus rugulipennis
L. wagneri
}   *Lygus maritimus* (common on weeds) differs in having few or no hairs on the upper surface, no coarse spines on tibiae and a complete keel or ridge between the eyes. It differs from *L. wagneri* in its pale green or yellow colour.

Lygocoris spp.   *L. contaminatus* differs from the three species in the key in having *brown* spines on tibiae arising from *black* spots.

Calocoris spp.   *Stenotus binotatus* differs in its yellow and black markings and by having tarsal segment 1 longer than segment 3. *Phytocoris ulmi* and *P. varipes* are brownish-red bugs with longer hind femora (longer than mid tibiae).

Orthonotus rufifrons.   *Mecomma ambulans* ♂ has all-black antennae. ♀ has black head.

VI.1. *Aphrophora* sp.

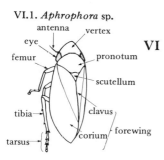

VI.2. *Cixius* sp.
(Fulgoromorpha)
VI.3. *Philaenus* sp.
(Cicadomorpha)

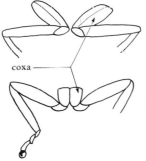

VI.4. (left) *Delphacodes* sp.
VI.5. (right) *Cixius* sp.

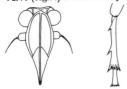

VI.6. (left) *Cixius* sp.
VI.7. (right) *Philaenus* sp.

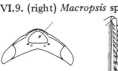

VI.8. (left) *Philaenus* sp.
VI.9. (right) *Macropsis* sp.

VI Homoptera (1): frog-hoppers, leaf-hoppers etc.

This is a very large group consisting mainly of small insects often differing only in fine details. Some of the characters given are supplementary to the couplets proper and are designed to distinguish the species from others *not* given in the keys but which might be encountered.

1 Basal segments (coxae) of middle pair of legs long and widely separated (VI.2). (Fulgoromorpha)   2

— Middle coxae short and set close together (VI.3). (Cicadomorpha)   4

2 Hind tibia with large leaf-like spur at its tip (VI.4). Lower margin of eye with a well-developed recess for the base of the antenna. (Delphacidae). Several species occasionally   e.g. *Javasella* spp.

— Hind tibia with several spines of roughly equal size at tip   3

3 Scutellum large, blackish, with 3 longitudinal keels (VI.5). Forewings transparent with small brown spots on veins. Face as in VI.6. ♂ 6.5—7.4 mm, ♀ 7.1—8.0 mm May—Oct   *Cixius nervosus*

— Not as above   Other fulgoromorph 'frog-hoppers'

4(1) Hind tibia cylindrical (without keels), with 1 or more fixed spines (VI.7). Vertex with roughly semi-circular plate (VI.8). (Cercopidae). (Species concerned about 6—10 mm)   5

— Hind tibia distinctly keeled and bearing rows of movable spines (each in a ball-and-socket joint) (VI.9). (Cicadellidae)   7

Cercopidae

5 Large, distinctive 'frog-hopper' with bold pattern of black and dark red markings on the wings. 9.5—10.5 mm Apr—July   *Cercopis vulnerata*

— Not as above   6

6 Larger species. Wings, pronotum and head covered with strong black dots. Pronotum and vertex with pale, raised median line. Forewings brownish with 2 distinct white patches on the outer margins (pl. 3.6). 9.1—10.2 mm May—Oct   *Aphrophora alni*

— Smaller species without strong black dots. Pronotum and vertex without raised keel. Usually straw-coloured but a range of colour forms occur with blackish-brown mottling or bands or even completely dark. Common June—Nov   'Cuckoo-spit' insect *Philaenus spumarius*

Cicadellidae

VI.10. *Macustus* sp.

VI.11 & 12. *Empoasca* sp.

VI.13. *Aphrodes* sp.

VI.14. *Aphrodes bicinctus* ♀

VI.15. *Macustus* sp.

VI.16. *Macropsis scutellata*

VI.17. *Macropsis scutellata*

VI.18. *Eupteryx* sp.

Vertex greatly extended and pointed as seen from above (VI.1), without black spots. Forehead more or less sharply angled in side view (VI.10)   8

— Forehead rounded from both dorsal and side views (VI.11, VI.12). Vertex often with black spots   9

8 Vertex differentiated from face by a narrow ridge around the rim (VI.13); with a pale, raised median line (VI.14). ♂ vertex and pronotum boldly marked with chestnut or blackish-brown leaving two transverse yellow bands. ♀ forebody and wings straw-coloured with dark mottling. ♂ 5.0—6.5 mm, ♀ 5.9—7.8 mm July—Sept   *Aphrodes bicinctus*

— Vertex without ridge (VI.10) or median line but with 2 dark transverse bands (sometimes faint) (VI.15). Wings dark brown with conspicuous pale veins. ♂ 4.7—5.6 mm, ♀ 5.0—6.2 mm Apr—Sept   *Macustus grisescens*

9(7) Without any dark markings. Wings pale shiny translucent greenish, fading to yellow after death. Vertex, face and legs likewise greenish fading to yellow. 3.5—4.1 mm Aug—Nov—H—Mar—June   *Empoasca decipiens*

— Not entirely green or yellow   10

10 Only a narrow rim of the back of the head (vertex) visible from above (VI.16); pronotum extending forwards beyond front margin of eyes (VI.17). (Forewings clear or faintly brownish with dark brown veins. Scutellum yellowish with 2 black triangles). ♂ 4.4—4.6 mm, ♀ 5.2—5.5 mm July—Oct   *Macropsis scutellata**

— Vertex at least as long (from front to back) as length of eye; often with bold dark spots   11

11 Forewings often brightly patterned with yellow, brown and black markings; distinctly longer than body. Veins of forewings and hind wings as in VI.18   12

— Forewings pale yellow, sometimes with smoky-brown patches in the middle; not or little longer than body. Veins as in VI.19 (sometimes difficult to see)   15

VI.19. *Macrosteles* sp.

VI.20. *Eupteryx urticae*

VI.21. *Eupteryx cyclops*

VI.22. *Macrosteles variatus*

VI.23. *Macrosteles sexnotatus*

12 Rear edge of vertex with a median black mark (sometimes fused with the large pair of spots on the vertex). Face with a dark spot or spots at the top. Smaller, less brightly coloured species   13

— No median black spot at rear margin of vertex and without dark spots at the top of the face. Conspicuous orange-yellow species with brown and black markings 14

13 Black triangle at base of vertex wider than long (VI.20). A pair of black spots at top of face, usually well separated. (Wings with extensive brown and buff patterning. Pronotum and scutellum with dark pattern as in VI.20.) Very common. 2.9—3.6 mm
    June—Aug & Aug—Nov   *Eupteryx urticae**

— Black triangle at base of vertex longer than wide; sometimes very narrow and sometimes joined to the pair of spots in the form of a Y (VI.21). The top of the face usually with a large spot but sometimes two separate small spots or none. (Wings and thorax similar to *E. urticae*.) 3.3—3.7 mm
    June—Aug & Aug—Nov   *Eupteryx cyclops**

14(12) A bright yellow 'saddle' across the closed wings, interrupted by a pair of black spots meeting in the middle. Vertex with one pair of large, irregularly round black spots (pl. 3.5). Very common. 3.5—4.3 mm
    June—Aug & Aug—Nov   *Eupteryx aurata*

— Not as above
    Other *Eupteryx* species from Dead-nettle and other herbs

15(11) Vertex with a pair of black spots near hind edge well separated from a pair of larger black spots at junction with face (VI.22). Face clear yellow, central area (frontoclypeus) sometimes edged with black. 4.0—4.5 mm
    June—Oct   *Macrosteles variatus*

— Vertex with a pair of hammer-shaped spots between the 2 pairs of round black spots (VI.23). Face with a few incomplete transverse lines broken in the middle. 3.2—3.8 mm
    May—Oct   *Macrosteles sexnotatus*

See:
Le Quesne (1960—9)

VII   Homoptera (2): plant lice

VII.1. *Microlophium carnosum*

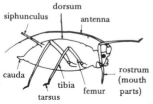

dorsum

siphunculus   antenna

cauda   rostrum (mouth parts)

tibia   femur

tarsus

VII.2. *Aphis urticata*

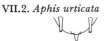

VII.3. *Aphis* sp

VII.4. *Myzus cymballariae*

1 Greenfly/blackfly — aphids. Plump soft-bodied insects with a pair of conspicuous tubes (siphunculi) on the abdomen (VII.1). Usually in colonies and often wingless (apterae). (Only the wingless forms of the 8 species known to feed on nettles are keyed out here and of these only *A. urticata* and *M. carnosum* occur regularly. A wide range of vagrant winged species may settle temporarily)   2

— Psyllids. Small, narrow, winged insects without siphunculi, but with a pair of prominent dark brown cones on the forehead (I.9). Jump when touched. Very common. 3.5 mm (pl. 3.7)
Oct—H—Jan—Mar(May)   *Trioza urticae**

2 Short antennae not reaching bases of siphunculi. Short siphunculi little longer than 'tail' (cauda) and less than distance between their bases (VII.2). Front of head more or less flat or gently curving (VII.3)   3

— Long antennae reaching beyond bases of siphunculi and often longer than body. Longer siphunculi, at least twice as long as cauda and equal to or longer than distance between their bases. Front of head recessed between lateral lobes (VII.4)   5

3 Dark green to black aphids. Usually in dense colonies on stems. 1.6—3 mm   4

— Pale yellow aphid. Usually scattered on undersides of leaves. 1.0 mm
June—Aug (summer generation)
       Small nettle aphid *Aphis urticata**

4 Smaller species. Legs brownish, without strongly contrasted pigmentation. Colour in life darkish mottled-green. Common. 1.6—2.0 mm
May—June (spring generation)
       Small nettle aphid *Aphis urticata**

— Larger species. Lower half of hind femora and tips of tibiae and tarsi are dark, remainder of legs quite pale. Colour in life uniformly black or blackish-green. 2—3 mm
May—Oct   Bean aphid *Aphis fabae*

VII.5. *Rhopalosiphoninus latysiphon*

VII.6. *Myzus ascalonicus*

VII.7. *Microlophium carnosum*

5(2) Siphunculi shiny black, greatly and abruptly swollen for about half their length (VII.5). Upper surface of body (dorsum) very convex, largely black and shiny. Occurs in dark, damp places on rhizomes or stems. 2.2—2.6 mm Apr—Oct

Bulb and potato aphid *Rhopalosiphoninus latysiphon*

— Siphunculi not black, not or only very slightly and gradually swollen. Dorsum not black   6

6 Siphunculi slightly but distinctly swollen toward their tips (VII.6); about as long as the distance between their bases. Recess in front of head wider than deep (VII.4)   7

— Siphunculi very long and narrow, if swollen then only very inconspicuously so (VII.7). Recess in front of head as deep as or deeper than wide   8

7 Legs, antennae and siphunculi all dark brown. (Siphunculi only slightly rough.) Colour in life pale brownish. 1.7—2.0 mm

Jan—Dec   *Myzus ascalonicus*

— Tips of tibiae dark, otherwise legs, antennae and siphunculi pale. (Siphunculi with rougher surface.) Colour in life dull yellowish-green or dark purplish-brown. 1.7—2.0 mm

Jan—Dec   *Myzus cymbalariae*

VII.8. *Phorodon humuli*

8(6) Front of head, and basal joint of antennae, with very conspicuous projecting lobes (VII.8). Colour in life greenish-white. 1.3—2.0 mm

June—Oct   Hop aphid *Phorodon humuli*

— Without such projecting lobes on the head   9

VII.9. *Myzus persicae*

VII.10. *Microlophium carnosum*

9 Smaller species. Side lobes of head broad and slightly converging (VII.9). Siphunculi very slightly swollen, with dark tips. Legs pale except for dark tarsi. Colour in life dull green. 2.0—2.5 mm

Jan—Dec   Peach potato aphid *Myzus persicae*

— Larger species (VII.1). Side lobes of head narrow, diverging (VII.10). Siphunculi not at all swollen, without dark tips. Tips of tibiae as well as tarsi dark. Colour in life pale green or dirty reddish. Common. 3.3—3.8 mm

May—Oct   Large nettle aphid *Microlophium carnosum**

See:
Blackman (1975)

VIII Free-living Diptera, Coleoptera and Neuroptera larvae: young stages of flies, beetles and lacewings

VIII.1. *Chrysopa* sp.

1 With three pairs of conspicuous, jointed legs and a well-formed head   2

— Fly maggots without true legs or head capsule   4

2 With long projecting jaws and antennae. Body often conspicuously hairy (VIII.1)
Lacewing larvae Chrysopidae

— Jaws not visible from above and antennae shorter than head   3

VIII.2. *Coccinella 11-punctata*

3 Body broad and tapering with rows of dark spots. Legs spreading (VIII.2). Up to 10 mm
Ladybird larvae Coccinellidae

— Body slender and pale with a pair of processes at the hind end (cerci). Head with 6 ocelli in a narrow ring (VIII.3). Up to 6 mm. (Staphylinidae)
Rove-beetle larvae *Tachyporus* spp.

4(1) Small pink maggots tapered at both ends. Up to 3 mm long   Cecidomyiid midge larvae *Aphidoletes* spp.

— Larger leech-like maggots. Usually green or brown with pale stripes, bars or warty tubercles. Up to 16 mm long (VIII.4)   Hoverfly larvae Syrphidae

VIII.3. *Tachyporus* sp.

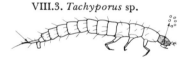

All of these are predatory on aphids, small caterpillars and other insects and can be reared to the adult stage if kept well supplied with food. The ladybirds may then be identified from key IV.

VIII.4. *Syrphus* sp.

See:
Dixon (1960)
van Emden (1949)

PLATE 1

1
Polygonia c-album
(Comma)

2
Pleuroptya ruralis
(Mother-of-pearl)

3
Abrostola triplasia
(Spectacle)

4
Phragmatobia fuliginosa
(Ruby tiger)

5
Anthophila fabriciana
(Nettle tap)

6
Hypena proboscidalis
(Snout)

7
Diachrysia chrysitis
(Burnished brass)

8
Phlogophora meticulosa
(Angle shades)

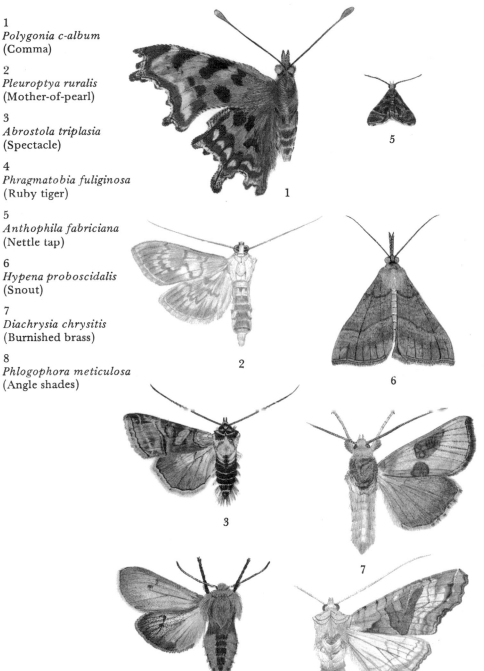

10 mm

PLATE 2

1
Dasineura urticae galls
(Nettle midge)

2
Orthosia gothica larva
(Hebrew character)

3
Melanchra persicariae larva
(Dot)

4
Polygonia c-album larva
(Comma)

5
Diachrysia chrysitis larva
(Burnished brass)

6
Abrostola triplasia larva
(Spectacle)

7
Hypena proboscidalis larva
(Snout)

8
Spilosoma luteum larva
(Buff ermine)

9
Pleuroptya ruralis leaf roll
(Mother-of-pearl)

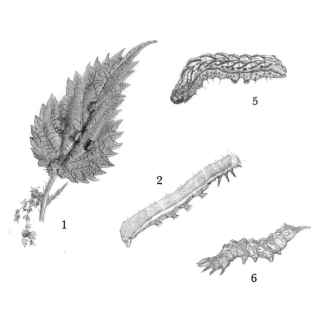

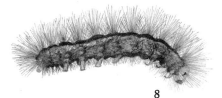

10 mm

PLATE 3

1
Phyllobius pomaceus

2
Ceutorhynchus pollinarius

3
Brachypterus urticae

4
Demetrias atricapillus

5
Eupteryx aurata

6
Aphrophora alni

7
Trioza urticae

8
Scathophaga stercoraria
(Yellow dung fly)

5

1

6

2

7

3

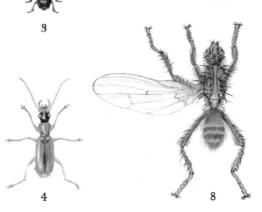

4

8

10 mm

PLATE 4

1
Heterotoma planicornis

2
Orthonotus rufifrons ♀

3
Liocoris tripustulatus
(Common nettle capsid)

4
Heterogaster urticae
(Nettle ground bug)

5
Aptus mirmicoides
(Ant damsel bug)

6
Lygocoris pabulinus
(Common green capsid)

7
Deraeocoris ruber

1

5

2

3

6

4

7

10 mm

PLATE 5

1
Propylea 14-punctata
Fourteen-spot ladybird

2
Rhagonycha fulva
Soldier beetle

3
Pholidoptera griseoaptera
Dark bush-cricket

4
Chrysopa carnea
Green lacewing

5
Phryxe nemea

6
Anthocoris nemorum

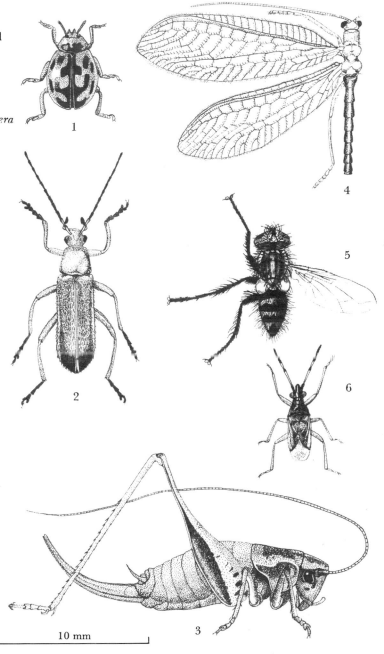

10 mm

IX   Diptera: mines and galls

IX.1. *Agromyza anthracina*

A. LARVAE

1 Swollen galls on leaf blades, stalks or flowers (pl. 2.1)
(Cecidomyiidae)
June—Aug/Sept   Nettle midge *Dasineura urticae**

— Mines in leaves or stems (Agromyzidae)   2

2 Leaf-miners producing conspicuous mines   3

— Stem-borers not visible externally; larvae pupate in
stems   5

IX.2. *Agromyza anthracina*

spiracles

3 Winding mine usually confined between two veins
initially and not adjoining margin of leaf; ends in a large
blotch (IX.1). (Spiracles of larva each with 3 bulbs on a
conspicuous bump) (IX.2)
June—July & Sept—Oct   *Agromyza anthracina**

— Irregular linear or blotch mine, the first part normally
following the margin of the leaf (IX.3). (Spiracles of
larva as above)   4

IX.3. *Agromyza reptans*

4 Spiracles of larva each with 4 groups of hairs
June—July & Sept—Oct   *Agromyza pseudoreptans**

— Spiracles of larva without hairs
June—July & Sept—Oct   *Agromyza reptans**

5(2) Larva with spiracles separated by a distance equal to
their own diameter and each consisting of an oval slit
with about 13 bulbs around a strong central horn
*Melanagromyza aenea**

— Spiracles of larva with up to 3 irregular bulbs and no
central horn   *Phytomyza flavicornis**

B. FLIES REARED FROM MINES AND GALLS

1 Gall midge. Very small fly with long antennae. (Under a
microscope an antenna looks like a string of beads
carrying whorls of short hairs)
Nettle midge *Dasineura urticae**

— Miners. Larger flies with short antennae (IX.4)   2

2 Stem-miners   3

— Leaf-miners   4

IX.4. *Agromyza anthracina*

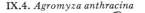
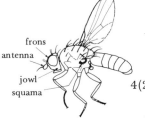

frons

antenna

jowl

squama

IX.5. *Agromyza pseudoreptans*

IX.6. *Agromyza reptans*

See:
Darlington (1968)
Spencer (1972)

3 Smaller delicate species with yellowish sides, legs, frons and antennae. Wing 2.1–2.6 mm
Phytomyza flavicornis *

— Larger, robust species. Black with brilliant-green to greenish-black abdomen. Wing 2.7–3.1 mm
Melanagromyza aenea *

4(2) Frons pale (reddish). Fringe of hairs on squama black (IX.4). Wing 2.4–3.3 mm   *Agromyza anthracina* *

— Frons matt black or dark brown. Fringe of hairs on squama pale. Wing 2.8–3.9 mm   5

5 Jowls narrow, at rear only about one ninth vertical height of eye (IX.5)   *Agromyza pesudoreptans* *

— Jowls deeper at rear, one quarter to one sixth vertical height of eye (IX.6)   *Agromyza reptans* *

Other flies

In addition to these plant-feeding flies, a wide variety of other flies will be found sheltering or hunting among nettles. These are too numerous and their association with nettles too casual to be included in the keys but some of the most common predators and parasites are described on pp. 42–3.

X   Parasites

A simplified key to the more common families likely to be bred from insects on nettles (see p. 44).

NB: Many Hymenoptera are extremely small and delicate and may best be examined in alcohol, or under a compound microscope. Their antennae are sometimes divided into distinct sections (X.9) but all segments should be counted. Females are recognised by their protruding ovipositor ('sting'). Males cannot always be identified with certainty.

X.1. *Aphelopus* sp.

1   Flies: insects with only 1 pair of wings. Antennae short, with a broad lobe and terminal hair. (IX.4). DIPTERA   2

— Insects with 2 pairs of wings, often very delicate. Bodies often slender and with conspicuous waist. Antennae of at least 5 segments (usually more), often threadlike or beaded. PARASITIC HYMENOPTERA   3

X.2. Larval dryinid attached to *Eupteryx* sp. (wings removed)

2   Small, narrow-bodied flies (2.5–5 mm) with outstandingly large spherical head. Parasites of Homoptera
                      Pipunculidae

— Larger, broad and conspicuously bristly flies (about 4–8 mm). Head not noticeably large. Parasites of caterpillars and a few Heteroptera (pl. 5.5)   Tachinidae

X.3. Ichneumon from caterpillar of *Abrostola triplasia*

3(1) Adults with antennae of 10 segments. Hind wings with posterior lobe (X.1). Larva is an external parasite within a brown or black sac attached to nymphal or adult leafhoppers (X.2)   Dryinidae

— Not as above   4

4   Antennae of 14 or more segments. Forewings with obvious veins forming cells and a dark patch (pterostigma) on front edge (X.3). Ichneumon flies   5

— Antennae of 5 to 13 segments. Veins of forewings much reduced or absent   6

5   Forewings with cross vein (X.3)   Ichneumonidae

— Forewings without cross vein (X.4). (Includes very small aphid parasites)   Braconidae

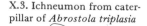

X.4. *Microgaster* sp.

6(4) ♀ ovipositor visible only from extreme tip of abdomen; abdomen often margined at sides. (Scelionoidea)   7

— ♀ ovipositor showing just before tip of abdomen, its line

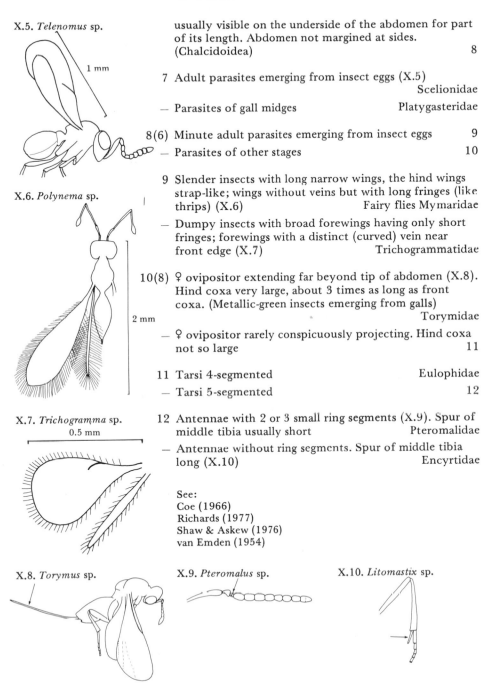

X.5. *Telenomus* sp.

1 mm

X.6. *Polynema* sp.

2 mm

X.7. *Trichogramma* sp.

0.5 mm

usually visible on the underside of the abdomen for part of its length. Abdomen not margined at sides. (Chalcidoidea)　　　8

7　Adult parasites emerging from insect eggs (X.5)
　　　　　　　　　　　　　　　　　　　Scelionidae

—　Parasites of gall midges　　　　Platygasteridae

8(6)　Minute adult parasites emerging from insect eggs　　9

—　Parasites of other stages　　　　　　　　10

9　Slender insects with long narrow wings, the hind wings strap-like; wings without veins but with long fringes (like thrips) (X.6)　　　Fairy flies Mymaridae

—　Dumpy insects with broad forewings having only short fringes; forewings with a distinct (curved) vein near front edge (X.7)　　　Trichogrammatidae

10(8)　♀ ovipositor extending far beyond tip of abdomen (X.8). Hind coxa very large, about 3 times as long as front coxa. (Metallic-green insects emerging from galls)
　　　　　　　　　　　　　　　　　　　Torymidae

—　♀ ovipositor rarely conspicuously projecting. Hind coxa not so large　　　　11

11　Tarsi 4-segmented　　　　　Eulophidae

—　Tarsi 5-segmented　　　　　　　12

12　Antennae with 2 or 3 small ring segments (X.9). Spur of middle tibia usually short　　Pteromalidae

—　Antennae without ring segments. Spur of middle tibia long (X.10)　　　　Encyrtidae

See:
Coe (1966)
Richards (1977)
Shaw & Askew (1976)
van Emden (1954)

X.8. *Torymus* sp.　　　X.9. *Pteromalus* sp.　　　X.10. *Litomastix* sp.

3 Biology

The keys cover about 100 species or groups of species associated with Stinging nettles. The selection is a somewhat arbitrary one for it includes some quite uncommon or local insects which are known or thought to depend on nettles (e.g. the weevil *Apion urticarium*) as well as many widespread species which live perfectly well on other plants but which are *commonly* found on nettles (e.g. the capsid bug *Plagiognathus arbustorum*). The kind of association differs greatly and there are various ways in which the insects can be subdivided ecologically. One obvious and important way is by their feeding habits; another way is by the times of year when they are found. Each species has evolved a special and highly tuned life style to make use of the resources provided by nettles and to compete successfully with all the other species that share similar needs: life is like a game in which every species must win continuously or become extinct. Most insects have very short lives; in the case of the nettle fauna the maximum life span is a little over one year and in many species each generation lasts only a few months. In this brief time they must therefore eat, grow up, mate, and reproduce successfully, avoiding predators, parasites and natural and other disasters. The chances of survival vary greatly from place to place and from one time to another and different species have evolved a great variety of strategies to ensure success somewhere. We know quite a lot about the life histories of some species but there is much more that we do not know. There is a theory that species which compete for the same resource cannot coexist unless they differ in some ecological respect. Thus when two species appear to have the same habits, it is a fair indication that we simply do not know enough about their biology. They may favour different parts of the country, different soils or degrees of shade; and even if they share precisely the same needs, one may be better at avoiding predators while the other produces more offspring.

Ecological separation of the three leaf-hoppers of the genus *Eupteryx* which live on nettles has been at least partially worked out recently by Le Quesne (1972) and Stiling (1980*a*, *c*). All three species have two generations a year and overwinter as eggs. However, *E. urticae* and *E. cyclops* are virtually confined to nettles (*E. urticae* occurs also on Pellitory-of-the-wall) whereas the second-generation eggs of *E. aurata* are laid on a wide range of unrelated plants such as Hogweed *Heracleum sphon-dylium*, Mint *Mentha rotundifolia* and Ragwort *Senecio jacobaea* in addition to Stinging nettle. In one particular study, the eggs laid on nettles by both generations of all three species of *Eupteryx* were found to be heavily attacked by parasites but the eggs of *E. aurata* on Hogweed and Hemp agrimony *Eupatorium cannabinum* were not attacked. It therefore looks as though this dispersal to other plants by *E. aurata* is a stratagem to reduce parasitism. The distinction between *E. urticae* and *E. cyclops* is less evident but *E. cyclops* may favour damper habitats.

A study of the leaf-mining and stem-boring agromyzid flies on nettles may likewise reveal subtle but important differences.

Feeding habits
Plant-feeders

The insects in this book can be divided broadly into those that feed on plants and those that feed on other insects, mites, etc. The former are termed phytophagous whilst the latter include predators and parasites, and scavengers that eat dead animal material. (This division is not quite clear-cut because a few species, mainly plant bugs (Heteroptera), are partially predatory.) The phytophagous species can themselves be subdivided in various ways. Table 2 shows how they can be classified according to the different parts of nettle plants that they use. There is a distinction not only between groups of insects but, in some cases, between larval and adult stages. This is, of course, most marked in caterpillars, leaf-miners, etc., in which the adult moths and flies do not feed on nettles at all. Next we have groups like the weevils that use different parts of the plant as larvae and adults, whereas plant bugs and leaf-hoppers show more subtle if any differences in their requirements. The distinction is lost completely in aphids during periods of parthenogenetic (asexual) reproduction, in which wingless females produce live young and both adults and juveniles live together in mixed colonies. On the other hand, the Small nettle aphid *Aphis urticata* shows different feeding habits at different times of year, and the two distinct forms were originally described as separate species. The spring generation is dark green or black and forms dense colonies on the stems of nettle. The summer generation is much less commonly noticed as it is smaller, pale yellow and lives in scattered colonies on the undersides of leaves, where it feeds from the main veins.

Taken together, the insects found on Stinging nettle provide a good example of how all parts of the plant are 'shared out'. They also demonstrate many types of vegetarian feeding behaviour — biting and chewing by caterpillars and adult weevils, mining by fly and weevil larvae, gall-forming by midge larvae and sap-sucking by the Hemiptera. The sap-suckers may be further divided into those that tap the sap-transporting vessels of the plant (phloem), such as the aphids, frog-hopper larvae (cuckoo-spit insects) and psyllid larvae, and those, including the plant bugs, leaf-hoppers and thrips, which merely suck the contents of individual cells and thus cause pale stippling on the leaves.

How do such relatively large and soft-bodied creatures as caterpillars avoid being stung? The claspers of the pro-legs and the underside of the body would appear to be particularly vulnerable and it would be worth making careful observations under a hand lens or microscope. The mouthparts are hard enough to chew up most of the

Table 2. *Parts of nettle plants used by different insects for feeding*

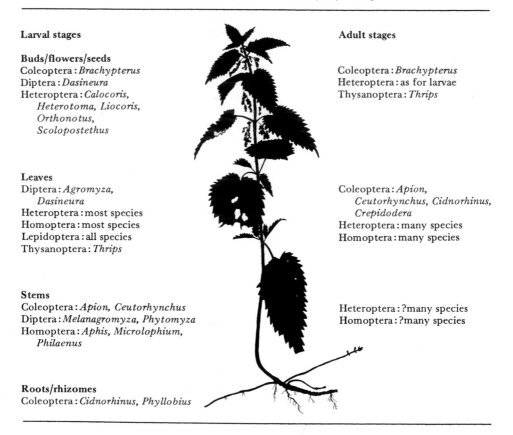

Larval stages

Buds/flowers/seeds
Coleoptera: *Brachypterus*
Diptera: *Dasineura*
Heteroptera: *Calocoris,*
   Heterotoma, Liocoris,
   Orthonotus,
   Scolopostethus

Leaves
Diptera: *Agromyza,*
   Dasineura
Heteroptera: most species
Homoptera: most species
Lepidoptera: all species
Thysanoptera: *Thrips*

Stems
Coleoptera: *Apion, Ceutorhynchus*
Diptera: *Melanagromyza, Phytomyza*
Homoptera: *Aphis, Microlophium,*
   Philaenus

Roots/rhizomes
Coleoptera: *Cidnorhinus, Phyllobius*

Adult stages

Coleoptera: *Brachypterus*
Heteroptera: as for larvae
Thysanoptera: *Thrips*

Coleoptera: *Apion,*
   Ceutorhynchus, Cidnorhinus,
   Crepidodera
Heteroptera: many species
Homoptera: many species

Heteroptera: ?many species
Homoptera: ?many species

hairs but in fact these are often bitten off at the base; if a few Small tortoiseshell larvae are kept in a glass jar with some leaves one can soon see an accumulation of stinging hairs lying at the bottom. (See Thurston & Lersten, 1969, for an account of the chemistry and functioning of stinging hairs in *Urtica*.)

Another useful distinction can be made between species that are restricted to Stinging nettle or at least to the genus *Urtica*, those that are limited to close relatives of nettle in the family Urticaceae or the order Urticales (see Introduction), and thirdly polyphagous (literally 'many-feeding') species that feed on nettles and unrelated plants such as grasses or thistles or Dead-nettles. Insects in the first two categories are asterisked in the keys and are summarised in table 3. Some of these associations were found by a simple survey of Pellitory-of-the-wall by Davis & Lawrence (1974) and more extensive studies on this species or wild Hop may reveal other insects that share this habit.

Predators and scavengers

Table 4 lists the most important predators on aphids and the eggs and young stages of plant bugs, hoppers and caterpillars; several others may occasionally attack other insects. The plant bugs and fly larvae have piercing and sucking mouthparts most evident in large nabid bugs. The beetles and remaining groups have chewing mouthparts whose rapacious nature is best displayed by lacewing larvae (VIII.1) but also clearly seen in the curved

Table 3. *Insects virtually restricted to nettles, or to one or more members of the Urticales in addition to* Urtica *itself*

Coleoptera
 Weevils *Apion urticarium, Ceutorhynchus pollinarius, Cidnorhinus quadrimaculatus* P, *Phyllobius pomaceus,* flower beetle *Brachypterus urticae* P
Diptera
 Leaf-miners *Agromyza anthracina* P, *A. pseudoreptans, A. reptans,* stem-borers *Melanagromyza aenea, Phytomyza flavicornis,* gall midge *Dasineura urticae*
Heteroptera
 Nettle ground bug *Heterogaster urticae* P, Common nettle capsid *Liocoris tripustulatus* P, *Orthonotus rufifrons*

Homoptera
 Leaf-hoppers *Eupteryx cyclops, E. urticae* P, *Macropsis scutellata,* aphids *Aphis urticata, Microlophium carnosum* P, psyllid *Trioza urticae* P
Lepidoptera
 Nettle-tap *Anthophila fabriciana* P, Mother-of-pearl *Pleuroptya ruralis,* Small tortoiseshell *Aglais urticae,* Peacock *Inachis io* H, Comma *Polygonia c-album* HE, Red admiral *Vanessa atalanta* PH, Burnished brass *Diachrysia chrysitis,* Snout *Hypena proboscidalis* H, Spectacle *Abrostola triplasia,* Dark spectacle *A. trigemina* H
Thysanoptera
 Thrips urticae

P, also on Pellitory-of-the-wall; H, also on Hop; E, also on Elms.

jaws of soldier beetles and the common little *Demetrias atricapillus* (pls. 5.2 and 3.4).

Scorpion flies and some of the beetles (*Crepidodera, Demetrias, Rhagonycha*) only come onto nettles when they are adult (the larvae live in the soil), whereas the hoverflies are only predatory in the larval stages. However, most of the species listed, including the five ladybirds and all the plant bugs, are to be found on nettles during both larval and adult stages. Scorpion flies are largely scavengers on dead animals, the Dark bush-cricket *Pholidoptera griseoaptera* (pl. 5.3) feeds on both plants and insects, and earwigs are pretty well omnivorous. None of these insects is confined to nettles as such but they favour the kind of tall dense vegetation that nettle beds provide. Earwigs are largely nocturnal and, like many of the noctuid caterpillars, often conceal themselves at the base of the plants during the day. Their presence and influence is therefore easily underestimated. Very little is known about the activities of most insects at night and there is a wide-open field of 'nocturnal ecology' waiting to be explored.

In addition to these insects, there are various flying predators that hunt among rank vegetation. One very common and conspicuous species is the Yellow dung-fly *Scathophaga stercoraria* which occurs almost all the year round in the vicinity of cattle-grazed pastures (pl. 3.8). This can often be seen and caught with another fly — perhaps a leaf-miner — held firmly in its legs. Other common predatory fly families include robber flies (Asilidae) and long-headed flies (Dolichopodidae). Social and solitary wasps may also be encountered, but these are outside the scope of this book, as are spiders (see Appendix). Together, these various predators must take

Table 4. *The principal insect predators on Stinging nettles*

Beetles: Coleoptera	Plant bugs: Heteroptera	Flies: Diptera
Coccinella 7-punctata	*Aptus mirmicoides*	Syrphidae, several L
Coccinella 11-punctata	*Dolichonabis limbatus*	*Aphidoletes* sp. L
Adalia bipunctata	*Nabis rugosus*	
Adalia 10-punctata	*Anthocoris* spp.	Lacewings: Neuroptera
Propylea 14-punctata	*Orius* spp.	*Chrysopa* spp.
Tachyporus spp.	*Deraeocoris ruber*	
Demetrias atricapillus A	(*Dicyphus errans*)	Earwigs: Dermaptera
Rhagonycha fulva A	(*Orthotylus ochrotrichus*)	*Forficula auricularia*
	(*Heterotoma planicornis*)	
Scorpion flies: Mecoptera	(*Calocoris sexguttatus*)	Crickets: Orthoptera
Panorpa spp. A	(*Liocoris tripustulatus*)	*Pholidoptera griseoaptera*

A, adults only; L, larvae only. Species in brackets are occasionally predatory.

a heavy toll of the many plant-feeders; otherwise even nettles would soon succumb to their onslaught.

It is very difficult to assess the various mortality factors that operate under natural conditions on a given prey species but some idea of the absolute or relative importance of various predators may be gained by observing and recording their activities. A study by Perrin (1976) on the Large nettle aphid *Microlophium carnosum* found anthocorid bugs to be the most numerous predators, followed by spiders, ladybirds and other plant bugs with smaller numbers of cecidomyid midge larvae, hoverfly larvae, soldier beetles, lacewings and earwigs. Banks (1955) showed that nettle aphids were an important prey for the first generation of several species of ladybirds which subsequently migrated onto field beans to attack the aphids there. One must not ignore bird predators. Whitethroats are known to favour nettle beds and even Great tits may transfer their attentions from trees to nettles and other herbs as a source of caterpillars to feed their later broods (Royama, 1970).

Parasites

Parasites tend to be considered a specialist interest as they are often very small and difficult to identify. However, any life history studies are likely to involve parasitism and it is easier to measure the effects of parasitism than predation in quantitative work even if one cannot name the parasites precisely. Likewise, anyone who simply wishes to breed out caterpillars to the adult stage will sooner or later find that his hoped-for Peacock butterfly or Burnished brass moth has given place to a black hairy fly or a small wasp-like creature. There are times, indeed, when one may despair of ever rearing a particular species successfully. This is not too surprising when one realises that parasitism is one of the chief ways in which insect populations are kept in check and that the prevalence of parasitism in a species tends to rise as its own numbers rise. Conversely, the avoidance of parasitism must be one of the chief aims of a target species and is one of the main reasons for dispersal to new sites (see the comment on *Eupteryx aurata* made in the last section). Small tortoiseshell butterflies have been seen to 'select' some experimentally grown nettle beds in the middle of a corn field for laying several batches of eggs when much larger areas of nettles around the edge of the field were apparently ignored.

Table 5 lists the more common and distinctive groups of parasites that may be encountered in studies of insects on nettles. The flies (Diptera) are represented by two families, of which the Tachinidae contains many widely polyphagous species such as *Phryxe nemea* (pl. 5.5). This species attacks all the Nymphalidae on nettles — Comma, Peacock, etc. — many of the Noctuidae such as the Dot moth, and the Pyralidae — Small magpie and Mother-of-pearl. Smaller species are known to attack plant bugs. The Pipunculidae only attack Homoptera and will be met as parasites of frog-hoppers such as the 'Cuckoo-spit' insect *Philaenus spumarius*, and leaf-hoppers *Eupteryx* spp. Small as pipunculids are, if you watch nettle beds carefully you may notice them rivalling the larger hoverflies (Syrphidae) in their ability to hover and manoeuvre in confined spaces. You may even be able to observe one land on the back of an unsuspecting hopper to lay an egg (see Colyer & Hammond, 1968).

The order Hymenoptera contains a very large number of parasites divided among many families of which 11 are given here. The ichneumon-flies (Ichneumonidae) contain the largest parasites and there is usually only one per caterpillar, which emerges and spins a cocoon often marked with light and dark bands. An example which emerged from a caterpillar of the Spectacle is illustrated in X.3. A similar group, the Braconidae, attacks a wide range of caterpillars on nettles including Red admiral,

Table 5. *Some of the principal genera of parasites known to attack insects on nettles*

(a) Parasites of: Parasite group	LEPIDOPTERA (caterpillars)
HYMENOPTERA	
Ichneumonidae	*Campoletis Diadegma Hyposoter Lissonota Netelia Oiorhinus Ophion Phobocampe Pimpla Stenichneumon* and many others
Braconidae	*Apanteles ˌAscogaster Clinocentrus Macrocentrus*[a] *Meteorus Microgaster*
Eulophidae	*Elachertus Eulophus*
Encyrtidae	*Litomastix*
Pteromalidae	*Pteromalus Dibrachys*
Scelionidae	*Telenomus* (in eggs)
Trichogrammatidae	? (in eggs)
DIPTERA	
Tachinidae	*Carcelia Ernestia Meigenia Pelatachina Phryxe Pseudoperichaeta Zenillia* and many others

(b) Parasites of:	HETEROPTERA	HOMOPTERA	DIPTERA	
			Miners	Gall midge
HYMENOPTERA				
Braconidae	?*Leiophron*	*Aphidius*	*Dacnusa Exotela*	
Mymaridae	*Polynema*	?		
Dryinidae		*Aphelopus*		
Torymidae				*Torymus*
Platygasteridae				?
DIPTERA				
Tachinidae	?			
Pipunculidae		*Chalarus Verrallia*		

? = genus not known.
[a]*M. grandis* is more or less specific to nettles though attacks a variety of hosts.

Autographa spp. and Garden tiger. They also parasitise leaf-mining fly larvae and several capsid bugs including *Liocoris tripustulatus*, *Plagiognathus arbustorum*, *Calocoris* spp. and *Lygocoris spinolai.*

An important subfamily of the Braconidae is the Aphidiinae, which attacks aphids. Brown or straw-coloured aphids are often very common amongst colonies of living insects, the coloration denoting the presence of an *Aphidius* larva within them. A bloated but empty skin with a circular hole is likewise conspicuous evidence of the emergence of an *Aphidius* adult. The genus *Trioxis* makes the aphid turn black, whereas *Praon* crawls out of its host before pupating and uses the dead skin of its victim like a bell-tent to protect its own cocoon.

Two families, the Torymidae and Platygasteridae, will be found attacking the Nettle gall midge *Dasineura urticae*, whilst the Mymaridae and Scelionidae are tiny parasites on the eggs of several groups of insects. Two thirds of the eggs laid on nettles by leaf-hoppers of the genus *Eupteryx* are sometimes affected by mymarids (Stiling, 1980*a*) and detailed studies on other Heteroptera and Homoptera would almost certainly show that many species are attacked at the egg stage. Of those *Eupteryx* that survive to the adult stage, a third or more may be affected by dryinid parasites, particularly in the second generation during August and September (Davis, 1973). Parasitised individuals can be readily recognised in the later stages by the sac-like larvae attached to the body (X.2). Males and females appear to be attacked equally and of course they are unable to breed. These two parasites together, acting successively on 67% and 33% of a population of *Eupteryx urticae*, would result in only 22 adults surviving out of every 100 eggs laid, assuming no mortality at all from other parasites, predators or other harmful influences.

Jervis (1980) has recently made a careful study of the dryinid and pipunculid parasites associated with several species of leaf-hoppers. However, much work remains to be done in associating parasites and their hosts in all groups except perhaps the larger Lepidoptera. Specialist taxonomists in the British Museum (Natural History) and elsewhere are generally too busy to identify the odd few specimens but might be interested to look at a well-documented series collected from known host species amongst say the Homoptera or Heteroptera. Results so obtained must be published, with acknowledgements, in a suitable journal so that the information is available to others.

Life cycles

The typical development of an insect is from a fertilised egg through a series of growing juvenile stages to the reproductive adult stage. The juvenile stages are all referred to as larvae in this book, but the term nymph is often applied to those insects in which the juvenile forms become increasingly like the adult form with each moult. This occurs, for example, in the plant bugs, hoppers, etc., in the orders Heteroptera and Homoptera and in the crickets and earwigs. The other insects such as the beetles, moths, flies, lacewings and parasitic Hymenoptera have larvae that are totally unlike the adult and they therefore have to pass through a pupal stage.

The larvae of the most common nettle-feeding plant bugs have been described in some detail by Southwood & Scudder (1956) and so it is possible, with a little effort, to identify both the species and the stage (instar). Fig. 2 illustrates the five larval stages of the Common flower-bug *Anthocoris nemorum*, showing the relative lengths of the second and third thoracic segments and the gradual development of wing buds. Fig. V.10 likewise shows a third instar larva of *Heterogaster urticae* which is the same size as fifth instar larvae of *Scolopostethus* sp. (V.9) or *Liocoris tripustulatus* (V.11). The immature stages of some Homoptera have been studied (e.g. woodland leaf-hoppers by Wilson, 1978) but there is considerable scope for extending Stiling's (1980*b*) study of the three *Eupteryx* species to other nettle-feeding species given in key VI. This would involve breeding specimens through from egg to adult, determining the periods spent in each instar and making careful notes, drawings and measurements of each, and finally deciding which characters are most useful for distinguishing each species.

Different species become adult at different times of year and survive in the adult stage for various lengths of time. The keys summarise the life histories of most of the insects found on nettles by giving the months when one can find the larvae of Lepidoptera and Diptera (miners and gall midges) and adults of the other groups. Fig. 3 illustrates some of the life cycle patterns found amongst nettle-feeding beetles, plant bugs and Homoptera in the East Midlands. (The timing and number of generations (phenology) may vary with many species in northern and western parts of the country but there is little definitive information on this.) The large green weevil *Phyllobius pomaceus* represents one extreme type of life cycle, with its single brief adult period in May/ June. During this time it is often a conspicuous insect in warm weather but after mating and laying eggs the adults

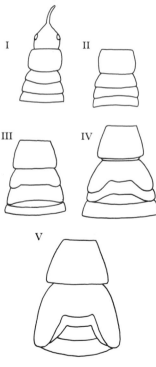

Fig. 2. The five larval stages of a plant bug *Anthocoris nemorum*.

I   II   III   IV   V

die and most of the year is spent in the larval stage underground. The small blackish weevil *Cidnorhinus quadrimaculatus* likewise has only a single generation a year but fresh adults first appear in September and October and reappear in the spring after hibernation. Some of these die after breeding but a proportion go into a summer resting phase (diapause) and re-emerge again in the autumn, perhaps to survive a second winter and breed a second year. There is no evidence that they breed more than once but why else would they live as long as twelve months and overlap with their own progeny? The same question applies to the related weevil *Ceutorhynchus pollinarius*. If enough adults could be

Fig. 3. Different types of life cycles seen in insects that feed on Stinging nettles. The top three are beetles, then come three capsid bugs, a leaf-hopper and the nettle psyllid. Note the different vertical scales indicating the relative abundance of adults of the various species in standard weekly samples and the big changes of scale in later generations of the bottom two species. Thickened lines indicate adults with eggs. Overwintering stages are shown by letters: A, adults; E, eggs; L, larvae; P, pupae. (Adapted from Davis, 1973.)

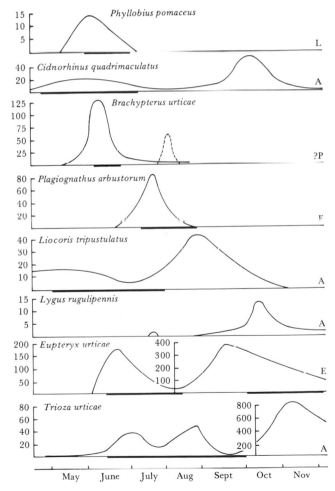

collected in the spring and kept in suitable conditions it
should be possible to determine the proportion that
survive diapause and to discover whether any do breed
again.

The flower beetles *Brachypterus* spp. probably have
two generations a year but under natural conditions
nettle flowers are only produced in quantity during quite
a brief period in June, so the second-generation larvae of
one or both species may depend mainly on other plants
(as in the case of the leaf-hopper *Eupteryx aurata* mentioned earlier); firm data on alternative host plants for
these two species are badly needed. Full-grown larvae
can be seen quite easily when they are nearly ready to
drop off the plant and pupate. They only occur on male
flowers as can be strikingly demonstrated by collecting
both male and female flowering stems from an area
where the adult beetles are seen, placing them in two
separate jars of water for a few days and counting the
numbers of larvae that drop off onto white trays.

Several capsid plant bugs have a single generation a
year with brief, synchronised adult populations in July,
about a month after *Phyllobius pomaceus*. These include
Plagiognathus arbustorum, *Orthonotus rufifrons*,
Heterotoma planicornis, *Orthotylus ochrotrichus*, *Lygocoris spinolai* and *Calocoris norvegicus*, all of which overwinter in the egg stage. The larvae of several of these can
be matched fairly easily with the adults and show a
similar wave-like pattern of increase and decrease as they
pass from egg to adult stages. The total number of individuals of the species concerned declines continuously of
course from the time the last egg is laid in August until
the next generation of eggs is laid the following year.
However, the eggs and youngest stages are usually difficult to count or estimate and one therefore receives the
very false impression that the species is most abundant
when it is actually near to its minimum for that year.

The common and nettle-specific capsid *Liocoris
tripustulatus* also has a single annual generation but
differs from the previous species by hibernating in the
adult stage and having a more protracted breeding period.
The overwintered adults tend to be much darker than
the young summer brood illustrated in pl. 4.3. The
European tarnished plant bug *Lygus rugulipennis* mirrors
Brachypterus in that a small early summer generation is
mainly passed on other plants whilst a larger second
generation is usually well represented on nettles. It can
be a pest on chrysanthemums, wheat and other crops.

The last two examples are taken from the Homoptera.
The very common leaf-hopper *Eupteryx urticae* has two
distinct adult peaks a year whilst the nettle psyllid
Trioza urticae probably has three or more generations a

year in the south of the country and fewer in northern
and upland districts. Populations build up from small
numbers of adults that survive the winter and lay eggs in
March, giving rise to a series of overlapping waves which
culminate in enormous populations in October.

Similar variations in life cycles can be seen amongst
the moths and butterflies (Lepidoptera), though here
one is concerned mainly with the larval stages. About
half of the species found on nettles hibernate as young
caterpillars and only one fifth have two generations a
year. The Angle shades *Phlogophora meticulosa* has
several overlapping broods and caterpillars can therefore
be found in every month of the year.

It is interesting to notice that the four butterflies
whose larvae feed on nettles, and — except for the Small
tortoiseshell — on their close relatives in the plant world,
are themselves members of the same family, the
Nymphalidae. A comparison of their life histories shows
important differences, which are summarised in table 6.
In all four species the adults hibernate but Red admirals
do not survive the winter in Britain and so we are
dependent on immigrant butterflies from the Continent
every year.

The potential rate of increase of most insects is
obviously very high, though little is known about this
in the case of the nettle fauna. Simple caging experiments with pairs of individuals during the early part of
their breeding season would provide novel information
on the numbers of eggs laid and on the choice of sites
— under the leaves, at the base of the leaf stalks, on or in
the stems, in the upper or lower parts of plants. Some
information on oviposition by plant bugs is given by
Southwood & Leston (1959).

Table 6. *Different strategies adopted by four species of nymphalid butterflies whose
larvae feed on nettles*

	Number of generations per year in Britain and months when larvae occur			
	One, complete (immigrant adults)[a]	One, complete	Partial second	Two, complete
Eggs laid singly	Red admiral (*Vanessa atalanta*) August/September	—	Comma (*Polygonia c-album*) June (and August)	—
Eggs laid in batches, young cater-pillars in a com-munal web		Peacock (*Inachis io*) July		Small tortoiseshell (*Aglais urticae*) June and August

[a]Immigrants arrive from June onwards and in good years produce an early summer generation of
larvae (June/July) in the south.

4 Techniques and approaches to original work

Whole books have been written on methods for collecting and studying insects and it is only possible here to draw attention to some of the techniques and topics which apply particularly to insects on nettles. Further information and ideas can be obtained from Oldroyd (1958), Tweedie (1971), Ford (1973) and Lewis & Taylor (1967). If this book stimulates your interest in entomology generally, you may like to consider joining a society such as the Amateur Entomologists' Society or the British Entomological and Natural History Society, both of which publish attractive journals with articles and notes including various mapping and recording schemes.

Collecting insects

There are four basic methods that can be used for collecting insects from nettles but whichever method is used it is advisable to be adequately clothed: long sleeves and stout trousers are indispensable, and rubber gloves, overtrousers and gumboots are useful for sustained contact with nettles. Some nettles sting more violently than others and a tube of anti-histamine cream is a helpful standby.

The first method of collecting is simply to search by eye. This is the best way of finding galls and leaf mines and also the various leaf-spinning caterpillars. Any folded, rolled or distorted leaf is worth examining, even those that are withered and hanging down, since these may contain one of the brown tortricid caterpillars that often eat partly through the leaf stalk. The webs made by the little Nettle-tap caterpillar *Anthophila fabriciana* and the neat cigar-like rolls of the Mother-of-pearl caterpillar *Pleuroptya ruralis* (pl. 2.9) are distinctive enough but careful observations on the way leaves have been rolled or spun together may lead to useful ways of distinguishing the other species concerned. Empty leaf rolls often harbour young and adult *Anthocoris* bugs but it is not certain whether they have attacked the original occupant or are merely sheltering here. Beware of being misled by caterpillars that are merely pupating within a folded leaf without actually feeding on nettle. In woodlands, particularly, one may find tree-feeding species that have dropped from above.

Large webs and evidence of concentrated feeding are a sure sign of colonies of Tortoiseshell or Peacock cater-

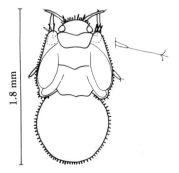

Fig. 4. *Trioza urticae* nymph and mouthparts.

1.8 mm

Fig. 5. Nettle leaves eaten by weevils: (*a*) *Cidnorhinus quadrimaculatus*, (*b*) *Phyllobius pomaceus*.

(*a*)

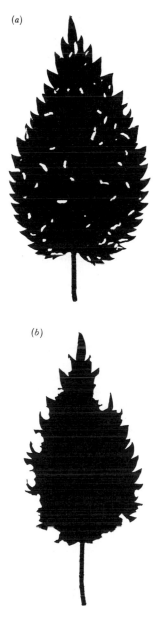

(*b*)

pillars, which tend to disperse quite widely in the later stages. Similarly, the presence of a colony of aphids is sometimes fairly evident, even from a distance, from the stunted and distorted leaves at the ends of the stems — and where these are present it is worth looking closely for ladybird, lacewing and hoverfly larvae and other aphid predators. Crinkled leaves are also produced by high densities of *Trioza* nymphs as a result of their sap-sucking activities. These are extremely small flattened creatures, quite unlike the adults, which adhere closely to the plant with their claws and long, sucking mouthparts (fig. 4).

Even holes in leaves can provide clues to the recent presence of free-living caterpillars and weevils. The two oval blackish weevils, for instance, make small round or elongated holes in the leaves and produce almost a lace-like effect where the feeding is concentrated, whilst the larger green *Phyllobius* weevil bites small jagged pieces out of the edge of a leaf (fig. 5). Similarly, third-stage caterpillars of the Snout *Hypena proboscidalis* have been found in late August making irregular holes in the blade of the leaf whereas full-grown specimens of this and other large caterpillars such as Garden tiger *Arctia caja* will consume the major part of a leaf.

The second and most productive method for mass collecting is to use a sweep-net made of stout cloth. The net is worked back and forth through the nettles vigorously and the insects collected into specimen tubes or pill boxes or sucked up into an aspirator ('pooter') as they fly or crawl up the net towards the opening. It may be helpful to stupefy the most active species by concentrating the catch in the bottom of the net and placing this in a wide-mouthed jar or polythene bag with a little ether or ethyl acetate for a few moments. The technique is particularly good for flying insects and can be made semi-quantitative by standardising the number of sweeps. However, it has limitations in dense nettle stands and is physically difficult where the nettles grow shoulder or head high. It is also less efficient for catching beetles and other insects which tend to drop to the ground when disturbed before the net can get underneath them. The *Cidnorhinus* and *Ceutorhynchus* weevils that are caught often remain motionless in the bottom of the net with their legs folded up for several minutes and until they move they are very difficult to spot.

The third method, which can be used for beetles, caterpillars and other groups that do not fly too readily, is a beating tray. This is placed gently next to a group of stems which is then knocked with a stick or shaken over the tray. This method again has limitations in dense nettle beds. Both the sweep net and beating tray should be tried at night time for collecting some of the noctuid caterpillars which often retire to the ground or to low vegetation during the day.

A good method for collecting the smaller and less active insects and for finding eggs is to place a number of leaves or whole stems into polythene bags and bring them home or to the laboratory for careful examination with a hand lens or binocular microscope. A drop of ethyl acetate on a wad of cotton wool may again be needed for stunning the insects temporarily, or killing them if necessary. When the collection is done carefully, this method can produce useful quantitative results for some of the bugs and hoppers (Heteroptera and Homoptera), for leaf-miners and even the more common caterpillars. If one suspects that stem-miners are present one could look for the larvae by splitting open the nettle stems. More information is certainly needed on the larval biology of these two agromyzid flies — their frequency, preferences, degree of parasitism etc. If you wish to make general deductions about the absolute numbers of any species in a nettle patch, or to make comparisons between species or between patches or between different occasions using any of these methods, you must pay careful attention to the sampling procedure. This is described in more detail later.

Two methods that are useful for special purposes are the emergence trap and the water trap. The first consists of a frame covered with fine muslin which can be placed over a small patch of nettles, either in the spring to trap insects emerging from their winter hibernation or later in the year to catch the emerging broods of adult leaf-hoppers, weevils, etc. The water trap uses a shallow, white or yellow tray containing a little water and a drop of detergent. Flying insects such as hoverflies are attracted to the bright surface and once wetted cannot get out.

Keeping live insects
It is always instructive to keep live insects. In the case of caterpillars and other immature stages it may be necessary to breed out the adult before the species can be identified with certainty. In other cases it may be desirable to record changes between moults or to observe feeding, egg laying or other behaviour. Rearing is quite easy with species like Small tortoiseshell *Aglais urticae* or Peacock *Inachis io* which pass through the whole cycle from egg to adult in a few weeks. They need only to be kept well supplied with fresh leaves in a clean cage

or jar with sufficient space for the chrysalis to hang
down and for the emerging butterfly to expand its wings
fully. Rearing is a longer and more risky operation when
pupation occurs in the autumn in moss, soil or among
dead leaves, as in the case of the Spectacle *Abrostola
triplasia* or Bright-line brown-eye *Lacanobia oleracea*.
With these species the pupae should be transferred to
clean containers and kept in a cool dry cellar or garage
during the winter and then watched carefully for
emergence in May or June. A little damp moss is useful
at this stage to provide humidity. Caterpillars that overwinter when half grown, like most of the Plusiinae and
hairy Arctiidae, are particularly difficult to maintain
but the same principles apply as for pupae. Both pupae
and larvae can be brought into a warmer environment in
the early spring to encourage them to emerge or recom-
mence feeding earlier than they would naturally. The
emerging adults must be provided with a twig or a
'ladder' of gauze so that they can climb up and let the
wings hang down and expand fully. The problem of
keeping overwintering larvae is avoided if these species
are sought in April or May, but by then winter has
taken its natural toll and relatively few may have sur-
vived. Care should be taken when introducing fresh
material that it does not harbour predators which
could attack and kill your caterpillars especially when
they are small. The eggs of parasites seen adhering to
the skin should also be carefully pricked with a needle
or squashed with a fine pair of forceps — unless one is
interested in rearing parasites of course!

Leaf-miners are not difficult to rear and too often the
mines themselves do not allow certain identification of
the species. However, the larva must complete its growth
within the one leaf and so this must be kept fresh. It is
best to choose the more advanced mines which only
require another day or two as the mines will remain
fresh if placed immediately and kept in a closed polythene bag. The emerging larvae should be placed in
individual tubes with a little fine, damp sand. Alterna-
tively, mined leaves may be placed directly on the surface of some damp sand in an earthenware flower-pot
covered with a sheet of glass. Excess moisture on the
glass should be removed periodically and the leaves
removed when pupation has taken place in the sand.
Adults of the first generation should emerge in three
to four weeks but the second generation will overwinter
as pupae and will need care to prevent them drying out.
This method is described in detail by Ford (1973). The
disadvantage of collecting older mines — and this applies
also to caterpillars — is that the risk of parasitism is

much increased. A useful technique for maintaining young *Agromyza* larvae under protected field conditions is to enclose the mined leaf in a small bag which can be conveniently made from a section of an old nylon stocking drawn together at each end.

The flower beetles *Brachypterus urticae* and *B. glaber* can likewise be readily bred out from nettle flowers towards the end of June by placing stems in water, catching the mature grubs on a white tray as they fall off and putting them in a pot of sand. The pupal stage lasts two to three weeks before new adults appear and disperse to other plants. The flower-pot method is also useful for rearing gall midges *Dasineura urticae*, though these have a longer larval period than miners and therefore it is even more necessary to take older stages in August or September so that the galls will not wither before the larvae emerge. Many interesting and useful observations could be made on miners, gall midges and flower beetles to establish their preferences and growth rates under different conditions.

Weevils pose no difficulties to keep as adults but their larvae live within the stems or rhizomes and so, like the two stem-mining flies, can only be maintained on living plants. Plants can be kept in large flower-pots and covered with nylon gauze stretched over wire hoops and the adult weevils can be marked in various ways to distinguish them from a new generation of emerging adults. Paint is liable to wear off after several weeks but careful scraping of scales from the underside of the body provides a permanent means of recognition.

Sap-sucking insects present a similar problem because some depend on actual sap pressure to obtain their meal. Plant-lice (aphids) and cuckoo-spit nymphs, for example, cannot be maintained on cut nettle stems and so these must be enclosed in gauze sleeves where they are found or transferred to potted nettle plants. Leaf-hoppers and heteropteran bugs do not need such forced feeding but they nevertheless require frequent fresh material. At least the older nymphs can be successfully reared on leaves kept within a pair of Petri dishes. Holes are drilled in the sides of the dishes so that the leaf stalks can be protruded and kept in tubes of water.

Several of the Heteroptera are partially or mainly predatory on aphids and other small insects. Like the various other predatory groups they require somewhat individual attention. The larger larval stages of nabid bugs, hoverflies, ladybirds, rove-beetles and lacewings should be fairly easily reared if one has access to a good supply of aphid-infested shoots.

Breeding from adult insects requires an additional

stage in that they need to be caged under suitable conditions for feeding, mating and choosing appropriate
egg-laying sites. For this purpose a tall muslin cage
should be placed over a patch of naturally growing
nettles or a vigorous potted nettle which has been prepared in advance.

Storing and displaying insects

All groups of nettle insects except moths and butterflies
can be stored for reference in 70% alcohol and examined
in solid watch glasses using a dropper pipette to transfer
small specimens. However, the adults of all but the
smallest and most delicate species are best mounted dry
and kept in cork-lined boxes with a repellant to ward off
attacks by mites etc. Naphthalene or paradichlorbenzene
crystals may be used as the repellant, or Rentokil
Woodworm Fluid painted around the edges of the box.
Traditional insect store boxes, made of wood and lined
with cork, are very expensive. A cheaper alternative can
be improvised by wedging a floor of expanded polyethylene foam (Plastazote; see p. 62) into a clear plastic
sandwich-box with a tight-fitting lid. Expanded poly-
styrene ceiling tiles make a cheaper (but much less satis-
factory) substitute for Plastazote.

Moths and butterflies and the larger flies etc. should
be pinned on setting boards with the wings either displayed conventionally (pls. 1.1 and 5.5) or in the natural
resting position (pls. 1.5 and 3.8) until they are set.

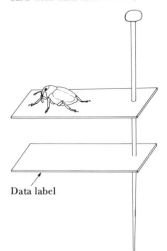

Fig. 6. Weevil mounted on
card with data label below.

Data label

Proper insect pins are essential, preferably stainless steel
pins as these do not rust. They can be obtained from
entomological dealers in different lengths and thick-
nesses to suit the insect and style of display. The flies
and smallest moths should be mounted with short fine
pins on strips of polyporus pith (or Plastazote trimmed
to the size and shape of a half matchstick) which are
themselves supported on long robust pins, as this makes
them easy to handle safely. The beetles, plant bugs, leaf-
hoppers, smaller flies and parasites, etc. are best
mounted on small rectangles or triangles of stiff white
card using a little gum and spreading out the legs and
antennae so that they can be clearly seen (pls. 3 and 4
and fig. 6). Powdered gum tragacanth with a drop of
water is often used by entomologists but wallpaper glue
or PVC gum can be used. (Note that you may need to
see the under-side of an *Adalia* ladybird or *Scolopostethus* plant bug first to identify it.) These cards are then
supported on large pins as above. To display the wings
and legs correctly it is important to relax the insects
properly before setting them. Freshly killed specimens
may require no special treatment but those that have

become stiff can be relaxed by placing them in a humid atmosphere in a tin or jar for twelve to twenty-four hours. Basic equipment for handling small insects includes a pair of delicate forceps, preferably stork-billed forceps, and a pair of curved, broad-ended, pinning forceps.

The smallest insects such as egg parasites, nettle thrips *Thrips urticae* or the young stages of the psyllid *Trioza urticae* need to be viewed through a compound microscope to see details of their structure. This involves mounting them on a glass slide in lactic acid or some more permanent medium such as polyvinyl lactophenol, but the techniques for this are beyond the scope of this book.

All specimens that are kept for reference, whether in alcohol or mounted for display, must carry data labels with the place and date of capture. (Use Indian ink or dark pencil for labels in alcohol.) A name label can be added when the specimen is identified but this can be done at any time. If the data label is forgotten or lost, the specimen is worthless as an ecological record. It may often be desirable to keep more detailed notes about a specimen than can be recorded on a data label; in this case each specimen should bear a code number *as well* and the details can be entered in a notebook against the appropriate serial number. This is especially useful, for example, where one has described an unknown caterpillar, perhaps through several moults, before it has finally emerged as a moth that can be named. There is unfortunately no easy way of keeping caterpillars except in alcohol, where they lose their natural coloration. Oldroyd (1958) described methods for emptying and blowing caterpillar skins, and for the extremely good freeze-drying technique, but the latter requires equipment which is only likely to be available to research institutions. Good, close-up colour photographs provide an excellent substitute.

A neatly set out collection of insects is both attractive and informative. For a study of insects on nettles one may wish to concentrate on a particular group like moths or plant bugs or to demonstrate the extreme range of insect types that are found. In either case it is well worth while spending time on careful mounting and labelling. Further details on this topic are given by Oldroyd (1958) and Ford (1973).

Drawing and dissecting
Drawing an insect is in itself a useful exercise for it makes one examine structures closely and notice features which otherwise escape attention. Compare, for

example, the antennae or legs of a weevil, a plant bug, a leaf-hopper and a fly (leaf-miner); or open out and draw their wings. Notice how in each case there is a difference in the form and function of the forewings and hind wings, taken to an extreme degree in flies, where the hind pair are modified into drumstick-like balancing organs (halteres). Examine also the clever folding of a beetle's hind wings that allows them to be fully protected by the elytra. Wing veins are frequently used in the identification of insects within large taxonomic groups such as flies, ichneumon-flies and leaf-hoppers and have been largely avoided in the keys here only because other, easier features can be used for the limited range of species concerned.

In order to draw small insects accurately and to scale, it is useful to view them with a microscope that has a grid graticule in one of the eyepieces and to use a sheet of paper which has been lightly ruled into squares. A partial substitute for the graticule is to draw a fine grid on a piece of white card or transparent celluloid and to place this underneath the object that is to be drawn. A more sophisticated aid is the drawing tube or camera lucida which can be attached to some microscopes. This allows the image of the object to be drawn around very easily.

It is sometimes necessary to dissect an insect to determine its breeding condition or its egg-laying capacity. For example, a maximum of 36 eggs has been found by dissecting a collection of the large green weevil *Phyllobius pomaceus* whilst the little nettle psyllid *Trioza urticae* has been found to lay an average of 176 eggs per female! Dissecting a small insect demands a good binocular microscope and a strong light, one or two pairs of Swiss watchmakers' forceps with very fine points and a mounted needle. The operation should be done under alcohol in a solid watch glass.

A soft-bodied insect such as a plant bug can be opened down the back of the abdomen fairly easily after opening or removing the wings. On the other hand, the little, hard-cased weevils are difficult to hold still and may have to be lodged in a small crater in a piece of plasticine. Fig. 7 illustrates the male and female reproductive organs of *Ceutorhynchus pollinarius*. Fig. 7(*b*) shows the immature condition of the ovary with its two pairs of slender milky white ovarioles seen in September shortly after the adult has emerged from the pupal stage. Fig. 7(*c*) shows an ovary containing developing eggs in November and fig. 7(*d*) an ovary which has shed its eggs as shown by the development of corpora lutea (literally 'yellow bodies') at the base of the oviducts.

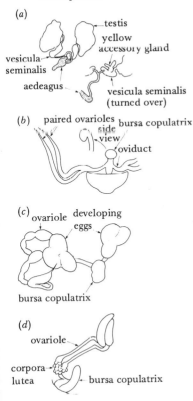

Fig. 7. Reproductive organs of the weevil *Ceutorhynchus pollinarius*: (*a*) male, (*b*)–(*d*) females at three stages of development.

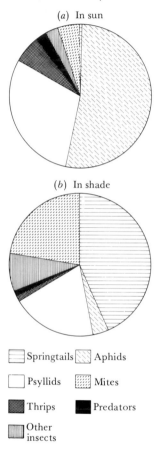

Fig. 8. The proportions of different groups of invertebrates collected on nettles (*a*) in the sun and (*b*) in the shade. (Redrawn from data by Wladimirsky (1926) after Uvarov (1931), with kind permission from the Royal Entomological Society of London.)

(*a*) In sun

(*b*) In shade

Springtails   Aphids

Psyllids   Mites

Thrips   Predators

Other insects

In each case only one side is fully shown. With care it is possible to observe the presence of eggs even in the tiny flower beetles *Brachypterus* spp. Dissection will also sometimes reveal internal parasites, for example in plant bugs.

Sampling and experimenting

A casual glance at Stinging nettles growing in open, sunny conditions and in well-shaded conditions will reveal obvious differences in the height of plants and the size of their leaves. Nettles growing on different soils or at different altitudes may vary in many other ways such as the density of plant stems (i.e. the number per square metre), the amount of flower and seed production or the density of stinging hairs on the leaves. Comparisons of this sort lead to a closer insight into the biology of Stinging nettle and its insect fauna and should prompt further enquiries. Fig. 8 shows the results of a simple study made in 1926 on nettle fauna in shaded and unshaded conditions.

If you want to measure such differences and draw general conclusions, it is important to base them upon proper sampling methods. First you must define what you want to measure; for example, plant growth might be characterised by leaf size, and leaf size could be determined by measuring its length, or length × breadth, or weight. You must then decide on a method of collecting a *representative* sample. It is almost impossible to prevent some personal bias in the collecting; even a small patch of nettles contains a great many leaves of differing sizes and one person may tend to favour lower leaves or those that are easier to reach (compare several collections made in this way by two people). You must therefore decide in advance that you will collect, say, 100 leaves selecting the largest one from every tenth plant or from the nearest plant at every tenth pace. If you can roughly measure the extent of the patch concerned you could divide it up in a suitable way (e.g. on a grid) and use a true random sampling system based on random numbers. These are provided by Lewis & Taylor (1967) and are also produced by many modern pocket calculators.

When the appropriate measurements or counts have been made at two contrasting sites, it is easy to calculate the average (mean) leaf size, plant height or density, etc., for each sample and see how big the differences are. If the two sets of figures overlap only a little, there is probably a real difference between your samples. However, if there is a big overlap, the difference between the mean values may be due just to chance and a statistical analysis is needed to decide whether the difference is sufficiently great to be 'significant'. (It is beyond the scope of this book to go into this but it is an important element of testing quantitative ecological theories.) If you wish to extend such a study to see if there are *consistent* differences between nettles growing under any contrasting conditions, the sampling process must be extended to cover several representative patches for each condition.

The same principles apply to collecting insects from nettles whenever one is concerned with making quantitative comparisons: whatever collecting method is used, it is important to carry out the sampling in the same systematic or properly randomised way at each site to avoid personal bias. However, insect numbers, or one's ability to catch them, may vary not only from place to place but from week to week and even from hour to hour, so it is even more important to define the question you are trying to answer and to standardise the sampling routine. How does temperature affect numbers caught in sweep net samples? Does the leaf-miner *Agromyza anthracina* favour shady conditions? What is the sex ratio of the frog-hopper *Aphrodes bicinctus* on nettles and does it vary with season? Are there consistent differences in colour forms of the cuckoo-spit frog-hopper *Philaenus spumarius* or nymphs of *Eupteryx urticae* in different habitats? (See Halkka, 1978; Stiling, 1980*b*.) Which aphid parasites or predators are most numerous in different places or at different seasons? Do the proportions of the two flower beetles *Brachypterus urticae* and *B. glaber* vary in different parts of the country? Methods for compiling and examining such data are given by Lewis & Taylor (1967).

An alternative approach to studying the biology of nettles and their insects is by doing experiments. This basically means imposing some treatment and examining the effects. The experiment could involve watering nettles or providing fertilisers or cutting down nettles at different periods of the year. How quickly do nettles regrow after being cut down in May or June or May *and* June and how does cutting affect their ability to produce flowers and seeds? What are the effects upon insects such as aphids or gall midges or flower beetles? Small tortoiseshell and Red admiral butterflies appear to favour lush young nettles to lay their eggs on in August/September. Can one produce this result experimentally? Again, it is necessary to treat two or more patches in the same way and to compare them with untreated patches to be sure that the results are consistent and not just a matter of chance.

Patches of nettle are generally close at hand and plentiful enough for sampling or maintaining livestock but they may not always be suitable for attracting par-

Fig. 9. Coding patterns for
marking up to 72 individuals
of *Phyllobius pomaceus* using
two colours separately and in
combination.

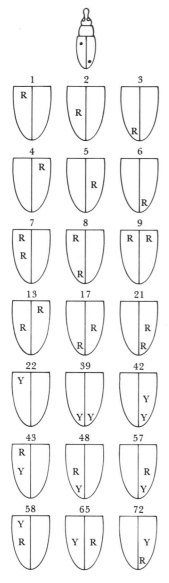

ticular insects or for controlled experiments. If you wish
to go further and create an experimental nettle bed, or
to grow nettles in pots which can be enclosed in cages or
moved about for different purposes, the quickest and
simplest method is to transplant whole plants or pieces
of rhizome. From autumn until early spring, fresh young
rhizomes can usually be found on or just below the surface of soil growing from an old rootstock. These are
ideal. They may well carry eggs or hibernating larvae,
pupae or adults, so if you are anxious to start with
insect-free material you should first shake off most of
the soil and dip the plants in a bucket of garden insecti-
cide such as dimethoate, fenitrothion or malathion.
Nettles do best in a rich loamy or peaty soil but almost
anything apart from pure sand or stiff clay will do. A
little high-phosphate fertiliser may be added and the
plants should be kept well watered if they are in pots.

Nettles can be grown quite easily from seed exposed
to light after they have been well soaked. Flowers and
rhizomes are not, however, produced in the first season
and seed-grown plants are generally much less robust
then those obtained from rhizomes.

Studies on the behaviour and life span of adult
insects can sometimes be made by marking individuals so
that they can be recognised again. The green weevil
Phyllobius pomaceus is a good subject for such studies
because it is large and has the habit of climbing up to the
tops of nettle stems. With a little practice, individuals
can be caught in 3 × 1 inch glass tubes and marked with
small spots of quick-drying cellulose paint on the elytra
using a pin or grass stem. A simple colour-coding system
can be used to distinguish different collecting dates or
even to identify separate individuals whose habits can
be recorded in a log book. Table 7 and fig. 9 show how
up to 72 weevils can be marked using only one or two
spots of red or yellow paint placed on the front, middle
or rear end of the two elytra. Repeated catches and
releases over a period of time will provide information
on movement of individuals, life span, sex ratios and

Table 7. *Number of different colour codings possible
using spots of paint on six distinct areas of an insect
body (see fig. 9)*

	No. of colours ▷ 1	2	3
No. of spots 1	6	12	18
2	15	60	135
Total (1 or 2)	21	72	153

means of estimating total populations of some colonies. Intensive sampling from small colonies may succeed in catching and marking all the individuals but more usually one will succeed only in marking a proportion of them, in which case the Lincoln Index formula can be used:

Total population = total number collected in second sample × number marked in first sample ÷ number of marked specimens recaptured.

For example, if 30 weevils are caught and marked in the first sample, and 40 weevils are caught in the second sample of which 10 are marked,

$$\text{Total estimated population} = \frac{40 \times 30}{10} = 120.$$

To be reliable, this formula depends on four things: (1) that the population remains the same from one sample to the other, i.e. there are no losses or gains, (2) that all individuals can be collected with equal likelihood, (3) that marked individuals are not more or less likely to be recaptured than unmarked ones, and (4) that a reasonably large number of marked weevils is caught in the second sample.

These requirements are best fulfilled by comparing catches over short intervals of time and by collecting and liberating insects over the whole nettle patch. Careful marking in this way does not appear to harm *Phyllobius* or affect its ability to fly but may render individuals more prone to predation by birds. Could this be tested?

How to present your findings

Writing up is an important part of a research project, particularly when the findings are to be communicated to other people. A really thorough, critical investigation that has established new information of general interest may be worth publishing if the animals on which it is based can be identified with certainty. Journals that publish short papers on insect biology include the *Entomologist's Monthly Magazine*, *Entomologist's Gazette*, *Bulletin of the Amateur Entomologists' Society*, and, for material with an educational slant, *Journal of Biological Education*. Those unfamiliar with publishing conventions are advised to examine current numbers of these journals to see what sort of thing they publish, and then to write a paper along similar lines, keeping it as short as is consistent with the presentation of enough information to establish the conclusions. It is then time to consult an appropriate expert who can give advice on whether and in what form the material might be published. It is an unbreakable convention of scientific publication that results are reported with scrupulous honesty. Hence it is

Some useful addresses

Suppliers of entomological equipment:

GBI (Labs) Ltd, Shepley Industrial Estate, Audenshaw, Manchester M34 5DW.
(Sole suppliers of Euparal; also general microscopic and entomological equipment, including chemicals.)

Watkins and Doncaster, Four Throws, Hawkhurst, Kent.

Worldwide Butterflies Ltd, Compton House, nr Sherborne, Dorset DT9 4QN.

Suppliers of expanded polyethylene foam (Plastazote, 12 mm thick):

Wilford Polyformes, Greaves Way, Stanbridge Road, Leighton Buzzard, Bedfordshire.

Suppliers of entomological books, new and second-hand:

E.W. Classey Ltd, PO Box 93, Faringdon, Oxon SN7 7DR.

Entomological Societies:

Amateur Entomologists' Society, 8 Heather Close, New Haw, Weybridge, Surrey KT15 3PF.

British Entomological and Natural History Society, c/o The Alpine Club, 74 South Audley Street, London W1Y 5FF.

Royal Entomological Society of London, 41 Queen's Gate, London SW7 5HU.
The Registrar of the Royal Entomological Society of London has a list of fellows of that Society who may be willing to advise on project work.

essential to keep detailed and accurate records throughout the investigation, and to distinguish in the write-up between certainty and probability, and between deduction and speculation. In many cases it will be necessary to apply appropriate statistical techniques to test the significance of the findings. A book such as Parker's (1973) *Introductory Statistics for Biology* will help, but this is an area where expert advice can contribute much to the planning, as well as the analysis, of the work.

Finding books

Some of the books and journals listed here will be unavailable in local and school libraries. It is possible to make arrangements to see or borrow such works by seeking permission to visit the library of a local university or by asking your local public library to borrow the work (or a photocopy of it) for you via the British Library, Lending Division. This may take several weeks. It is important to present your librarian with a reference that is correct in every detail. References are acceptable in the form given here, namely the author's name and date of publication, followed by (for a book) the title and publisher or (for a journal article) the title of the article, the journal title, the volume number, and the first and last pages of the article.

Appendix

1 Snails commonly associated with Stinging nettles
Arianta arbustorum (L.) Copse snail
Ashfordia granulata (Alder) Silky snail. Locally common
Cepaea hortensis (Müller) Garden snail
Cepaea nemoralis (L.) Grove (or Brown-lipped) snail
Monarcha cantiana (Montagu) Kentish snail
Trichia hispida (L.) Hairy snail
Trichia striolata (Pfeiffer) Strawberry snail. The most characteristic snail of nettle beds
See: Kerney & Cameron (1979) or Janus (1965)

2 Some spiders associated with nettle beds
(This is based on limited sampling and the list could undoubtedly be much expanded and annotated according to other habitat factors.)
Araneus cucurbitinus Clerck
Clubiona reclusa O.P.-Cambridge
Clubiona other spp.
Dictyna arundinacea (L.)
Enoplognatha ovata (Clerck)
Linyphia clathrata Sund.
Linyphia peltata Wider
Linyphia triangularis (Clerck)
Meta mengei (Blackwall)
Philodromus spp.
Tetragnatha montana Simon
Theridion pictum (Walck.)
Theridion sisyphium (Clerck)
Theridion other spp.
Xisticus cristatus (Clerck)

References and further reading

The *Handbooks for the Identification of British Insects* are published by the Royal Entomological Society of London, and can be bought from that society or from E.W. Classey Ltd (addresses on p. 62). Books marked with an asterisk are out of print.

Banks, C.J. (1955). An ecological study of Coccinellidae (Col.) associated with *Aphis fabae* Scop. on *Vicia faba*. *Bulletin of Entomological Research*, 46, 561–87.

*Beirne, B.P. (1952). *British Pyralid and Plume Moths*. London: Warne.

Blackman, R. (1975). *Aphids*. London: Ginn.

Carter, D.J. (1979). *Observer's Book of Caterpillars*. London: Warne.

Chandler, A.E.F. (1968). A preliminary key to the eggs of some of the commoner aphidophagous Syrphidae (Diptera) occurring in Britain. *Transactions of the Royal Entomological Society of London*, 120, 199–217.

Chinery, M. (1976). *A Field Guide to the Insects of Britain and Northern Europe*, 2nd edn. London: Collins.

Coe, R.L. (1966). Diptera. Pipunculidae. *Handbooks for the Identification of British Insects*, 10, no. 2(c).

Colyer, C.N. & Hammond, C.O. (1968). *Flies of the British Isles*, 2nd edn. London: Warne.

Darlington, A. (1968). *The Pocket Encyclopaedia of Plant Galls in Colour*. London: Blandford Press.

Davis, B.N.K. (1973). The Hemiptera and Coleoptera of Stinging nettle (*Urtica dioica* L.) in East Anglia. *Journal of Applied Ecology*, 10, 213–37.

Davis, B.N.K. (1975). The colonization of isolated patches of nettles (*Urtica dioica* L.) by insects. *Journal of Applied Ecology*, 12, 1–14.

Davis, B.N.K. & Lawrence, C.E. (1974). Insects collected from *Parietaria diffusa* Mert. & Koch and *Urtica urens* L. in Huntingdonshire. *Entomologist's Monthly Magazine*, 109, 252–4.

Dixon, T.J. (1960). Key to and descriptions of the third instar larvae of some species of Syrphidae (Diptera) occurring in Britain. *Transactions of the Royal Entomological Society of London*, 112, 345–79.

Ford, R.L.E. (1973). *Studying Insects: A Practical Guide*. London: Warne.

Greig-Smith, P. (1948). Biological flora of the British Isles: *Urtica* L. *Journal of Ecology*, 36, 339–55.

Halkka, O. (1978). Influence of spatial and host plant isolation on polymorphism in *Philaenus spumarius*. In *Diversity of Insect Faunas*, ed. L.A. Mound & N. Waloff, pp. 41–55. Oxford: Blackwell Scientific Publications.

Janus, H. (1965). *The Young Specialist looks at Land and Freshwater Molluscs*. London: Burke.

Jervis, M.A. (1980). Ecological studies on the parasite complex associated with typhlocybine leaf hoppers (Homoptera, Cicadellidae). *Ecological Entomology*, 5, 123–36.

Joy, N.H. (1932). *A Practical Handbook of British Beetles* (2 vols.). London: Witherby. (Reprinted in 1976 by E.W. Classey Ltd.)

Kerney, M.P. & Cameron, R.A.D. (1979). *A Field Guide to the Land Snails of Britain and North-West Europe*. London: Collins.

Kerrich, G.J., Hawksworth, D.L. & Sims, R.W. eds. (1978). *Key Works to the Fauna and Flora of the British Isles and North-western Europe*. London: Academic Press.

Kloet, G.S. & Hincks, W.D. (1964–78). Check list of British insects, 2nd edn. 1: Small orders and Hemiptera (1964). 2: Lepidoptera (1972). 3: Coleoptera and Strepsiptera (1977). 4: Hymenoptera (1978). 5: Diptera and Siphonaptera (1976). *Handbooks for the Identification of British Insects*, 11, nos. 1–5.

Le Quesne, W.J. (1960–9). Hemiptera, Cicadomorpha (1965); Hemiptera, Cicadomorpha (1969); Hemiptera, Fulgoromorpha (1960). *Handbooks for the Identification of British Insects*, 2, nos. 2(a), 2(b), 3.

Le Quesne, W.J. (1972). Studies on the coexistence of three species of *Eupteryx* (Hemiptera : Cicadellidae) on nettle. *Journal of Entomology A*, 47, 37–44.

Leston, D. & Scudder, G.G.E. (1956). A key to larvae of the families of British Hemiptera–Heteroptera. *Entomologist*, 89, 223–31.

Lewis, T. & Taylor, L.R. (1967). *Introduction to Experimental Ecology*. London: Academic Press.

Linssen, E.F. (1959). *Beetles of the British Isles* (2 vols.). London: Warne.

Oldroyd, H. (1958). *Collecting, Preserving and Studying Insects*. London: Hutchinson.

Parker, R.E. (1973). *Introductory Statistics for Biology* (Studies in Biology). London: Edward Arnold.

Paviour-Smith, K. & Whittaker, J.B. (1967). A key to the major groups of British free-living terrestrial invertebrates. In *The Teaching of Ecology*, ed. J.M. Lambert, *Symposium of the British Ecological Society* 7, pp. 24–32. Oxford: Blackwell Scientific Publications.

Perrin, R.M. (1976). The population dynamics of the Stinging nettle aphid, *Microlophium carnosum* (Bukt.). *Ecological Entomology*, 1, 31–40.

Richards, O.W. (1948). Insects and fungi associated with *Urtica*: insects. In P. Greig-Smith, Biological flora of the British Isles: *Urtica* L. *Journal of Ecology*, 36, 340–3.

Richards, O.W. (1977). Hymenoptera: introduction and keys to families, 2nd edn. *Handbooks for the Identification of British Insects*, 6, no. 1.

Royama, T. (1970). Factors governing the hunting behaviour and selection of food by the Great Tit (*Parus major* L.). *Journal of Animal Ecology*, 39, 619–59.

Shaw, M.R. & Askew, R.R. (1976). Parasites. In *The Moths and Butterflies of Great Britain and Ireland*, vol. 1, ed. J. Heath, pp. 24–56. London: Curwen.

South, R. (1963). *The Moths of the British Isles* (2 vols.), 4th edn. London: Warne.

Southwood, T.R.E. (1956). Key to determine the instar of an heteropterous larva. *Entomologist*, 89, 220–2.

Southwood, T.R.E. & Scudder, G.G.E. (1956). The immature stages of the Hemiptera–Heteroptera associated with the Stinging nettle (*Urtica dioica*). *Entomologist's Monthly Magazine*, 92, 313–25.

*Southwood, T.R.E. & Leston, D. (1959). *Land and Water Bugs of the British Isles*. London: Warne.

Spencer, K.A. (1972). Diptera. Agromyzidae. *Handbooks for the Identification of British Insects*, 10, no. 5(g).

Stiling, P.D. (1980a). Host plant specificity, oviposition behaviour and egg parasitism in some leaf-hoppers of the genus *Eupteryx* (Hemiptera : Cicadellidae). *Ecological Entomology*, 5, 79–85.

Stiling, P.D. (1980b). Colour polymorphism in nymphs of the genus *Eupteryx* (Hemiptera : Cicadellidae). *Ecological Entomology*, 5, 175–8.

Stiling, P.D. (1980c). Competition and coexistence among *Eupteryx* leafhoppers (Hemiptera : Cicadellidae) occurring on stinging nettles (*Urtica dioica*). *Journal of Animal Ecology*, 47, 793–805.

*Stokoe, W.J. & Stovin, G.H.Y. (1958). *The Caterpillars of British Moths Including the Eggs, Chrysalids and Food Plants* (2 vols.), 2nd edn. London: Warne.

Thurston, E.L. & Lersten, N.R. (1969). The morphology and toxicology of plant stinging hairs. *Botanical Review*, 35, 393–412.

Tweedie, M.W.F. (1971). *Pleasure from Insects*. Newton Abbot: David & Charles.

Uvarov, B.P. (1931). Insects and climate. *Transactions of the Royal Entomological Society of London*, 79, 1–247.

van Emden, F.I. (1949). Larvae of British beetles. VII. Coccinellidae. *Entomologist's Monthly Magazine*, 85, 265–83.

*van Emden, F.I. (1954). Diptera Cyclorrhapha. Calyptrata (1) Section (a). *Handbooks for the Identification of British Insects*, 10, no. 4(a).

Wilson, M.R. (1978). Descriptions and key to the genera of the nymphs of British woodland Typhlocybinae (Homoptera). *Systematic Entomology*, 3, 75–90.

Index

Abrostola, 18, 36, 41, 53
Adalia, 19—20, 42, 55
Aglais, 13, 14, 41, 49, 50—1, 52
Agromyza, 34, 35, 40, 41, 54, 59
Amphipyra, 17
Anthocoris, Anthocoridae, 22, 23, 42, 43, 46, 50
Anthophila, 12, 14, 41, 50
Aphelopus, 36, 44
Aphidius, Aphidiinae, 44, 45
Aphidoletes, 32, 42
aphids, 11, 12, 30—1, 39, 40, 41, 54
Aphis, 30, 39, 40, 41
Aphrodes, 28, 59
Aphrophora, 11, 27
Apion, Apionidae, 18—19, 38, 40, 41
Aptus, 22, 42
Arctia, Arctiidae, 16, 51, 53
Asilidae, 42
Autographa, 17, 45
Axylia, 18
beetles, 7, 10, 18—21, 32, 42
biology of insects, 38—49
Brachypterus, 20, 40, 41, 47, 48, 54, 58, 59
Braconidae, 36, 44
breeding and keeping of insects, 52—5
butterflies, 7, 49
Callimorpha, 16
Calocoris, 23, 24, 26, 40, 42, 45, 48
Cantharidae, 21
capsid bugs, 23—5, 48
Carabidae, 18
caterpillars, 12, 13—18, 49, 52—3
Cecidomyiidae, 32
Cercopis, Cercopidae, 27
Ceutorhynchus, 19, 40, 41, 47, 51, 57
Chalcidoidea, 37
Chrysopa, Chrysopidae, 32, 42
Cicadellidae, 27, 28
Cicadomorpha, 27
Cidnorhinus, 19, 40, 41, 47, 51
Cimicidae, 22, 26
Cixius, 27
Clepsis, 15
Coccinella, Coccinellidae, 19—20, 32, 42
Coleoptera, 5, 7, 10, 18—21, 32, 42
collecting methods, 50—2
Crepidodera, 20, 40, 42
crickets, 7, 10, 42
Curculionidae, 18—19
Cycnia, 16
Dasineura, 34, 40, 41, 45, 54
Delphacodes, Delphacidae, 27
Demetrias, 21, 42
Deraeocoris, 23, 26, 42
Dermaptera, 7, 10
Diachrysia, 16, 41
Dicyphus, 24, 26, 42
Diptera, 5, 7, 11, 32, 34—5, 36, 43
Dolichonabis, 22, 42
Dolichopodidae, 42
drawing and dissecting of insects, 56—8
Dryinidae, 36, 45

earwigs, 7, 10, 42
Elateridae, 21
Empoasca, 28
Encyrtidae, 37
Eugnorisma, 17
Eulophidae, 37, 44
Eupteryx, 28, 29, 38—9, 41, 45, 47, 48, 59; parasites of, 36, 44, 45
Eurrhypara, 13, 15
feeding habits of insects, 39—45
flies, 7, 11, 32, 34—5, 42
Forficula, 10, 42
frog-hoppers, 5, 11, 27—9, 40
Fulgoromorpha, 27
galls, 34, 40
Hemiptera, 5
Heterogaster, 23, 26, 41, 46
Heteroptera, 5, 7, 12, 21—6, 39, 40, 41, 42
Heterotoma, 25, 40, 42, 48
Homoptera, 5, 7, 11
hoverflies, 7, 11, 32, 42
Humulus, 5, 6, 41
Hymenoptera, 5, 44
Hypena, 13, 17, 41, 51
Ichneumonidae, 36, 44
Inachis, 5, 14, 41, 49, 50—1, 52
Javasella, 27
Lacanobia, 17, 53
lacewings, 7, 11, 32, 42
ladybirds, 19—20, 42
leaf-hoppers, 5, 11, 27—9, 48—9
Lepidoptera, 5, 7, 12—18, 46—9, 52—3
life cycles, 46—9
Lincoln index population estimates, 61
Liocoris, 9, 24, 26, 40, 41, 42, 45, 46, 47, 48
Litomastix, 37, 44
Lygaeidae, 22, 23, 25, 26
Lygocoris, 24, 26, 45, 48
Lygus, 23, 24, 26, 47, 48
Macropsis, 27, 28, 41
Macrosteles, 9, 11, 29
Macustus, 28
Mecomma, 26
Mecoptera, 7, 11
Melanagromyza, 34, 35, 40, 41
Melanchra, 17, 18
Microlophium, 12, 30, 31, 40, 41, 43
mines and miners, 34, 44, 53
Miridae, 21, 22, 23, 26
moths, 7, 10, 12—13
Mymaridae, 37, 45
Myzus, 30, 31
Nabis, Nabidae, 12, 21, 22, 26, 42
Neuroptera, 7, 11, 32
Noctuidae, 15, 16—18, 44
Nymphalidae, 44, 49
Olethreutes, 15
Orius, 21, 22, 42
Orthonotus, 25, 26, 40, 41, 48
Orthoptera, 7, 10
Orthosia, 12, 17
Orthotylus, 25, 26, 42, 48

Panorpa, 11, 42
parasites, 10, 36—7, 39, 43—9
Parietaria, 5, 6, 41
Philaenus, 27, 28, 40, 44, 59
Phlogophora, 17, 49
Pholidoptera, 10, 42
Phorodon, 31
Phragmatobia, 16
Phyllobius, 5, 19, 40, 41, 46—7, 48, 51, 57, 60—1
Phytocoris, 26
Phytomyza, 34, 35, 40, 41
Pipunculidae, 36, 44, 45
Plagiognathus, 25, 38, 45, 47, 48
plant bugs, 7, 12, 21—6, 39, 40, 41, 42
plant lice, 11, 30—1
Platygasteridae, 37, 45
Pleuroptya, 13, 15, 41, 50
Plusiinae, 53
Polygonia, 13, 41, 49
Praon, 45
predators, on insects on nettles, 41—3
Propylea, 20, 42
psyllids, 11, 30
Pteromalidae, 37, 44
Pyralidae, 14, 15, 44
Rhagonycha, 21, 42
Rhopalosiphoninus, 31
sampling and experimenting, 58—61
Scathophaga, 42
Scelionidae, 37, 45; Scelionoidea, 36
Scolopostethus, 22, 23, 25, 40, 46, 55
scorpion flies, 7, 11, 42
snails, associated with nettles, 62
spiders, associated with nettles, 62
Spilosoma, 16
Staphylinidae, 20
Stenotus, 26
storing and displaying insects, 55—6
Syrphus, Syrphidae, 11, 32, 42
Tachinidae, 36, 43—4
Tachyporus, 20, 32, 42
Telenomus, 37, 44
Thrips, 10, 40, 41, 56
Thysanoptera, 7, 10
Tortricidae, 14, 15
Torymidae, 37, 44, 45
Trichogramma, Trichogrammatidae, 37
Trioxis, 45
Trioza, 11, 30, 41, 47, 48, 50, 51, 56, 57
Udea, 15
Ulmus, 5, 6, 41
Urtica dioica, U. urens, 3—4, 6; as food for insects, 39—41; stinging hairs of, 40—1
Vanessa, 14, 41, 49
weevils, 18—19, 46—7, 51
writing-up, 61—2

Glossary

These terms are defined here in relation to the context in which they are used in this book. Many of the terms have a wider use and definition in other contexts.

abdomen Third major division of an insect body behind head and thorax

annual Term applied to plants that grow each year from seed

antenna 'Feelers' on the front of the head, varying greatly in form (IV.1, VI.1, VII.1, IX.4). Pl. *antennae*

anterior Front part, as opposed to posterior

aphid Plant louse (family Aphididae, order Homoptera) (VII.1)

apterae Wingless forms of aphids

asexual reproduction Reproduction without mating

brachypterous Short winged. Term applied to plant bugs, hoppers and beetles when wings are shorter than the abdomen (pl. 4.5 and V.2)

caterpillar Juvenile (larval) stage of moths and butterflies

cauda Tail-like process at the end of the abdomen in aphids (VII.1)

cerci Paired, jointed processes at the hind end of the abdomen

chrysalis Pupal stage of moths and butterflies

clavus Part of the rear portion of the forewing in Heteroptera and Homoptera, divided from the rest of the wing by a strong vein. The two clavi (pl.) meet behind the scutellum when the wings are folded together (V.1, VI.1)

commissure The straight line formed by the meeting of the clavi in Heteroptera (V.1)

connexivum The marginal area of the abdomen in Heteroptera

corium The main thickened area of the forewing in Heteroptera and Homoptera (V.1, VI.1)

corpora lutea Granular structures at the base of the oviducts, developed after eggs have been laid (fig. 7d)

costal fracture The break in the outer (front) edge of the fore-wings in Heteroptera (V.1)

coxa The basal segment of the legs, nearest the body (VI.2, VI.3). Pl. *coxae*

cuneus The tip of the thickened part of the forewings in Heteroptera beyond the costal fracture (V.1)

diapause A resting stage e.g. during the summer period

dorsum Upper surface of an insect (VII.1). Adj. *dorsal*

elytron Hard or leathery wing cases formed by forewings in beetles (IV.1). Pl. *elytra*

femur Third segment of an insect leg, often the longest or thickest (IV.1, VI.1, VII.1). Pl. *femora*

frons Forehead area between the eyes. Adj. *frontal*

frontoclypeus Central part of face of a Homoptera hopper

gall Swelling on leaf, stalk or flower bud caused by an insect (or mite)

genitalia Sexual structures

haltere Club-like balancing organ in flies (Diptera), replacing the hind wing (I.4)

hermaphrodite Containing both stamens and ovary

hibernate To survive in a dormant condition over winter

host Plant or larger insect that provides food or shelter

inflorescence Flowering stalk with flowers

jowl Lower part of the side of the face below the eyes in flies (IX.4)

keel A ridge, usually pronounced or sharp-edged, on head, legs, etc.

larva Juvenile stage of insects; term sometimes reserved for those that have a pupal stage before transition to the adult. Pl. *larvae*

lateral Side part or view

maggot Legless larval stage in flies

metamorphosis Change from pupal to adult state

micropterous With very short wings; term applied to some adult Heteroptera bugs

mine Tunnel made by an insect within the leaf or stem of a plant

nymph Larval stage in those insects that become increasingly like the adult with each moult

ocellus Simple eye, present singly or in small groups in some larval or adult insects (V.1, VIII.3). Pl. *ocelli*

omnivorous Feeding on both plant and animal matter

ovariole Subdivision of the ovary in which the eggs develop (fig. 7b)

ovary Reproductive organ which produces the eggs

oviduct Tube down which the ripe eggs are passed

oviposition Act or process of laying eggs

ovipositor Tube for inserting eggs into position; modified into a sting in parasitic Hymenoptera

parasite An insect whose larval stages feed on a larger, host insect; often internal and almost always causing the death of the host. (The more exact term is a parasitoid)

parthogenetic reproduction Reproduction by females without mating

perennial Plant that lives for several or many years

phytophagous Feeding on living plants

polyphagous Feeding on many kinds of food

posterior Hind part of body, etc.

prolegs Unjointed, fleshy legs at the rear half of a caterpillar; varying from two to five pairs (I.12)

pronotum First thoracic segment as seen from above; well developed in adult beetles, bugs and hoppers (IV.1, V.1, VI.1)

pubescence Covering of fine hairs

punctuation Fine pitting on the surface of an insect body or wing

pupa Stage between larva and adult during which metamorphosis occurs. Pl. *pupae*; vb. *pupate*

raptor Insect that catches prey. Adj. *raptorial*

rhizome Underground stem

rostrum Beak-like elongation of the head in weevils (IV.2, IV.3) or the jointed, sucking mouthparts of Heteroptera and Homoptera (I.8, I.10, I.11)

scutellum The large, triangular segment between the bases of the folded wings in Heteroptera and Homoptera (V.1, VI.1); very small in Coleoptera (IV.1). Adj. *scutellary*

seta Fine hair or bristle. Pl. *setae*

siphunculus Paired tube like processes on the abdomen of aphids (VII.1, VII.5, VII.6, VII.7). Pl. *siphunculi*

spiracle Breathing pore. Present in most insects but especially conspicuous as a row on each side in caterpillars and at the rear end of fly larvae (I.12, IX.2). Adj. *spiracular*

squama Circular or oval flap at the base of the wing in flies; often fringed with hairs and sometimes partly covering the halteres.

striation Fine line or groove, e.g. on elytra

tarsus Last section of an insect leg, usually subdivided into three to five short segments the last one bearing claws (IV.1, VI.1, VII.1). Pl. *tarsi*

thorax Major section of an insect body bearing three pairs of legs, and wings in the adult. Formed of three segments clearly seen in most larval stages (I.12, fig. 2)

tibia Fourth segment of an insect leg, often the longest (IV.1, VI.1, VII.1). Pl. *tibiae*

tubercle Small bump or short process

vertex The top of the head between and behind the eyes

viviparous Producing live young instead of laying eggs